AF737507

CIRCULAR ECONOMY

CIRCULAR ECONOMY

Multidisciplinary Approaches from Turkey

Edited by
NATALYA KETENCI

LEXINGTON BOOKS
Lanham • Boulder • New York • London

This book consists of studies presented at the conference organized by Yeditepe University Management Application and Research Center (YUVAM).

Published by Lexington Books
An imprint of The Rowman & Littlefield Publishing Group, Inc.
4501 Forbes Boulevard, Suite 200, Lanham, Maryland 20706
www.rowman.com

86-90 Paul Street, London EC2A 4NE

Copyright © 2023 by The Rowman & Littlefield Publishing Group, Inc.

All rights reserved. No part of this book may be reproduced in any form or by any electronic or mechanical means, including information storage and retrieval systems, without written permission from the publisher, except by a reviewer who may quote passages in a review.

British Library Cataloguing in Publication Information Available

Library of Congress Cataloging-in-Publication Data

Names: Ketenci, Natalya Shevchik, editor.
Title: Circular economy : multidisciplinary approaches from Turkey / edited by Natalya Ketenci.
Description: Lanham : Lexington Books, [2023] | Includes bibliographical references and index. | Summary: "This book reviews different industries and sectors that are becoming active players in the circular economy, on example of Turkey"— Provided by publisher.
Identifiers: LCCN 2022040533 (print) | LCCN 2022040534 (ebook) | ISBN 9781666921083 (cloth) | ISBN 9781666921090 (epub)
Subjects: LCSH: Sustainable development—Turkey. | Circular economy—Turkey. | Industries—Turkey.
Classification: LCC HC495.E5 C58 2023 (print) | LCC HC495.E5 (ebook) | DDC 338.9561—dc23/eng/20221107
LC record available at https://lccn.loc.gov/2022040533
LC ebook record available at https://lccn.loc.gov/2022040534

CONTENTS

Introduction

NATALYA KETENCI

The circular economy progressively is becoming a central part of our lives. Continuously growing population, increasing waste, environmental degradation, and depleting resources enforce economies to reconsider the conventional way of living. A circular economy focuses on the continuous use of resources without producing waste, so used products are becoming inputs for new products. The circular economy process ensures the sustainable use of scarce resources and improves environmental quality. Despite all its advantages the circular economy system is relatively new and breaks traditional ways of production. This is the process that requires new technologies and investments to restart conventional processes of production. Therefore, it is important to inquire about new knowledge in this area to share the experience between different sectors and countries. The circular economy gradually is being applied to all economic sectors, where new innovations allow to improve production efficiency with the minimum use of new resources.

The aim of this book is to analyze an experience and novelties in the circular economy system in Turkey. The circular economy system, functions, and outcomes are analyzed from different points of view in cases of different economic sectors of Turkey. One of the striking features of this book is that it combines different aspects of circular economy issues in Turkey and discusses possibilities for green economy developments to maintain sustainable growth.

This book consists of thirteen chapters, which can be organized into several parts. The first part of the book is devoted to an overview of sustainable supply chains, sustainable consumer behavior, and applications of the circular economy. This part reviews general concepts that are necessary for understanding the circular economy's functioning. The first chapter undertakes a systematic analysis of the literature on sustainable supply chains and circular economy intending to provide a thorough insight into the field employing bibliometric, network, and content analyses. This study identifies research gaps in the literature and offers future directions of the sustainable supply chains linked to the circular economy. The next chapter analyzes sustainable consumer behavior summarizing local and global research on sustainable consumption since 2017 and exploring if the sustainability efforts can be considered as a module in Customer Experience Journeys of companies. The final chapter of this part investigates and reports an extensive review of the current applications of circular economy concepts in practice in Turkey at both private sector levels such as manufacturing, textile, petrochemicals, electronics, construction, automotive, food, agriculture, etc.; and government policy level.

The next part is devoted to microeconomic issues of circular economics and to sectoral analysis. Chapter 4 examines how spare parts are revised according to product groups under the business model, whereas as a result of the economic benefits provided to the end-users, the benefits to the country's economy and the environment are revealed. This chapter discusses and provides answers to the following questions: How does this circular economy business model contribute to waste reduction and optimization of resource efficiency? How does it contribute to extended product life cycles in Turkey? The next chapter analyses e-waste recycling and explores the key antecedents of consumer intention toward recycling e-waste. The study uses the "warm glow" factor by adding to the original model of the theory of planned behavior, as one of the intrinsic motivation factors. Findings of the study support that the perceived behavioral control and the warm glow variables have a positive and significant effect on the e-waste recycling intention of consumers. However, the attitudes towards recycling and subjective norms don't have a significant effect on the behavioral intention towards recycling e-waste. Chapter 6 assesses consumer dynamics for green homes. The study tests the mediating effect of attitude towards green homes between independent variables (environmental concern, environmental awareness, subjective norm) and purchase intention. The study aims to shed light on future academic studies on consumer housing preferences and practically guide the marketing strategies of the developers and real estate agencies as well as nongovernmental organizations in the sector in promoting environmentally friendly residential buildings.

The next chapter of the microeconomic part examines sustainability and the transportation sector and attempts to answer the question if transportation can go green. Green transportation has subjects in greenhouse emissions, fuel consumption, and transportation efficiency. Because transportation is deemed as a sector that has high carbon emissions, it is valuable to mention its present and future situations. This study evaluates the sustainability of transportation and makes projections for the future. Chapter 8 is focusing on the energy sector in Turkey and examines how this sector is exposed to the circular economy. This study analyzes and discusses if there are any changes in the energy sector in Turkey towards the circular economy and if these changes have a noticeable impact on the green economy.

The next part of the book is focusing on the governmental issues of the circular economy. Chapter 9 explores the role of municipalities in the circulation of secondhand textile products. This study visualizes municipalities' practices in the circulation of secondhand textile products, in terms of their role in the process, motives, and challenges they face with. The study discusses three major findings: delineation of the whole process as of collection, sorting and cleaning, displaying and re-distribution, domination of social and economic motives, and various challenges in the system. The next chapter focuses on the financial aspects of environmental, social, and governance criteria in Turkey. This chapter evaluates the impact of environmental, social, and governance (ESG) criteria on the financial performance of firms quoted on the Borsa Istanbul Stock Exchange. The results of the study reveal that in Turkey, there is a significant and positive relationship between environmental, social, and governance criteria and the financial performance of the selected firms for the analysis period.

The next part of the book is analyzing the circular economy from the macroeconomic point of view. Chapter 11 focuses on the material flow analysis in Turkey. This chapter aims to measure circularity in Turkey at the macro economic level by analyzing the "circular material use rate (CMU)" as a relevant indicator over time. Findings of this study indicate that the degree of circularity in Turkey's economy measured as the share of recycled materials in overall processed materials appears to be quite low, 2.43 percent, 3.16 percent, and 4.56 percent in 2014, in 2016, and in 2018 respectively. Comparing Turkey's CMU with other European countries suggest that Turkey must revisit its policies and regulations to improve its circularity and decrease the demand for virgin raw materials as recycled material could replace them. Chapter 12 focuses on the indicators of a circular economy for Turkey at the macro level and addresses the potential economic benefits of converting from linear to circular models for Turkey. This study argues that the major trade-off for Turkey is that the need for sustainable economic growth leads to excessive environmental degradation since the two main drivers of the economy are the construction and energy sectors. Finally, the last chapter concludes with the analysis of a revolutionary COVID-19 recovery approach based on a circular economy. This research examines the presently existent circular economy approaches in Turkey responding to the pandemic and considers their potential effect on prosperity growth, job creation while reducing environmental pressure and being better prepared to save lives. On one hand, this chapter aims to summarize the direct and indirect efforts by Turkey towards establishing circular economy components through reusing various medical devices, altering hospitals' supply chains, recycling chemicals, and reducing air pollution to enhance public health. On the other hand, the chapter will demonstrate how policymakers can offer investment opportunities in various economic sectors post-pandemic by building a sustainable socio-economic model of development.

This book is the collection of studies that were presented at the 5th Yeditepe International Research Conference on Business Strategies entitled Circular Economy for a Sustainable Future: A Multidisciplinary Approach. The conference was organized by Management Applications and Research Center (YUVAM).

1

An Innovative Perspective on Supply Chains in the Context of Circular Economy
A Bibliometric Analysis

ZEHRA BINNUR AVUNDUK

Introduction

Global economic advancements have led to increasing demand for natural resources such as clean water, forests, metals, and crude oil (Lee and Tang, 2017). Nevertheless, natural resources are turned into waste from production to consumption in a linear economy. Waste production causes the environmental deterioration by reducing the value of natural capital caused by pollution from waste and by diminishing natural capital (through mining or unsustainable harvesting) from the environment. On the other hand, the circular economy (CE) is an approach that emphasizes the development of innovative ways to use natural resources more sustainably by focusing on environmental concerns due to resource scarcity and ecological degradation. CE represents the current movement to conceptualize the integration of environmental well-being and economic activity in a sustainable manner (Murray et al., 2015). Dissimilar to the conventional linear economic model based on a "take-make-consume-throw away" model, CE is based on reusing, sharing, repairing, leasing, refurbishing, remanufacturing, and recycling in an (almost) closed-loop. In practice, CE implies diminishing waste to the lowest level. Moreover, adopting the CE approach is vital to achieving the worldwide net-zero emissions target by 2050, set by the 2015 Paris Agreement.

Increasing resource scarcity causes rising raw material prices and price volatility, negatively impacting firms' profitability and nations' economic prosperity. This circumstance reshapes competitive precedence for businesses and supply chains worldwide (Ellen MacArthur Foundation, 2014;

Masi et al., 2017). Businesses, micro-units of the economy, use natural resources in all processes from production to consumption. However, every stage in the firms' operations reveals a transaction cost. One business provides inputs to other firms to produce in the value chain, while others transform these inputs into final goods or services. Firms work as a network at all stages, from material supply, production of goods and services, distribution, marketing, and delivery of products to end-users. Thus, the entire system of inter-firm activities forms supply chain networks, which are the meso-units of the economy. The transition from linear supply chains to sustainable ones shortens the stages and reduces transaction costs. CE advocates the recycling of energy and requisite materials as well as the capability for one firm's wastes to be used as a resource by another firm through self-organization competencies. The CE concept promotes sustainable development by preventing the creation of critical resources and environmental challenges. It also emphasizes that financial systems can perform according to the material and energy cycle principles that sustain natural systems (Zhu et al., 2011). Therefore, for the CE to reach its purpose, the traditional supply chain structure must be transformed into an innovative, dynamic, environmentally friendly, and lean form that prevents waste and sustainably use resources.

The CE concept's fundamental premise is that the value recovery from products through the closer cycles of refurbishment and reuse are superior both environmentally and economically to energy recovery and recycling. Therefore, facilitating such cycles, supply chains are presumed to be a crucial action element for implementing the CE because of the requisite for a combined effort (Masi et al., 2017). According to Zhu et al. (2010: 1325), *"beyond organizational boundaries to involve external coordination with upstream partners to obtain environmentally friendly inputs and with downstream partners to cooperate for environmental management practices such as product return, reuse, and recycling."* Nevertheless, innovative supply chains that permanently sustain the value circulation by integrating the diversity of possible CE—supply chain forms and are completely alignment to CE fundamentals have not been adequately addressed in academia and practice. Therefore, a structural analysis is required to explore changes in the structures of renewed supply chains and the new trends in the circular economy context. This study aims to answer these research queries using the methodology of content analysis with bibliographic coupling to spotlight the latest research trends and gaps and recommend future research proposals.

The remainder of this chapter is structured as follows: Section 2 presents the research methodology and describes BibExcel, Gephi, and

Leximancer used as bibliometric analysis software. Then, section 3 shows the findings of the research. The final part of the chapter provides a brief conclusion by discussing the results obtained from the study and suggestions for future studies.

Methodology of the Research

This study aims to analyze the content of articles and visualization of the literature on sustainable supply chain and CE published in the Web of Science Core Collection database by using BibExcel, Gephi, and Leximancer software. All articles that appeared in the search results were included in the research sample without any year limitation. The WoS database resulted in 239 articles on sustainable supply chains and CE. Summary, title, and references of the articles in the selected sample were compiled as the unit of analysis and examined with the specified programs.

According to Krippendorff (2019), "content analysis is a technique for making replicable and valid inferences from texts (or other meaningful matter to the context of their use." Content analysis is a research methodology that allows researchers to analyze and quantify certain concepts, words, or themes within some given qualitative data. Furthermore, it is a systematized analysis of texts that intends to find out themes, concepts, and relationships in the collection and unveil unexplored qualities about the data to generate trustworthy and valid inferences (Krippendorff, 2019). In addition, it extracts meaningful value from text data to understand better the research domain's content (Hofmann and Chisholm, 2016), and mapping the structure presents it visually. The entire purpose is to understand the characteristics and content of existing research in the field and provide direction for scholars and practitioners (Kumar et al., 2020).

This study conducted co-citation network and bibliographic coupling (BC) analysis (Kessler, 1963) to determine clusters and streams of the specified sample of the research. BC is one of the essential quantitative techniques for mapping the literature and examining the improvement of academic disciplines (Zupic and Čater, 2014). Bibliometric tools allow conducting a transparent, systematic, and replicable review to analyze extant research and provide a comprehensive knowledge map of the scientific area (e.g., Kraus et al., 2020; Maseda et al., 2022). In addition, bibliometric techniques have been used to map the scientific areas such as innovation (Gao et al., 2020; Liu et al., 2015; Randhawa et al., 2016; Shang et al., 2019; X. Zhang et al., 2016), supply chain management (e.g., el Baz and Iddik, 2022; Fahimnia et al., 2015; Maditati et al., 2018;

Montecchi et al., 2021; Zhang and Zhao, 2021; Zhang et al., 2020), CE (e.g., Goyal et al., 2021; Luis and Celma, 2020; Singh et al., 2021; Tsai et al., 2020) and other areas.

BC links papers that reference the identical set of cited papers and is used to evaluate the similarity among citing documents. This method is forward-looking and suitable for capturing current research trends and studying emergent literature fields within a research area (e.g., Boyack and Klavans, 2010; Vogel and Güttel, 2013). In addition, indicating the probability of a shared topic, BC associates documents with similarities in their reference lists (Maseda et al., 2022). Applying BC in this study pursued previous research, specifically Zupic and Čater (2014) and Randhawa et al. (2016). The coupling-based proximity values were initially calculated using the BibExcel packaged software (Persson et al., 2009). Then the proximity scores were used to examine, and the relation between the publications using network analysis was visualized. Network approaches have become progressively popular in bibliometric studies as they present a correct, effectual, and clear visualization of a larger size of texts in a meaningful way (Randhawa et al., 2016; Vogel and Güttel, 2013; Zupic and Čater, 2014). This study applied the "Force Atlas 3D" network algorithm implemented in the Gephi software to visualize the network (Bastian et al., 2009). In the second stage of the research, the "Louvain" modularity optimization technique was used to determine clusters (Blondel et al., 2008).

The Leximancer software performs a style of automatic content analysis. The program operates both relational (semantic) and conceptual (thematic) analyses of textual data (Rooney, 2005). It enables researchers to review the actual words of authors and, thus, determine themes and concepts emerging from literature (Biesenthal and Wilden, 2014). Leximancer assists in a text-driven and objective review of the literature. Since it is a machine learning algorithm, it selects the most proper number of themes relying on the explored concepts (Wilden et al., 2017). In addition, the Leximancer method seeks to explore indirect, implicit relationships among concepts. In this way, previously unknown relationships can be discovered (Smith and Humphreys, 2006).

An asymmetric scaling algorithm mapped the concepts in a semantic network. The algorithm assesses the proximity among concepts relying on how often they co-appear in the document and broadens beyond simply looking at concepts' appearances. In these meaning maps, "entity concepts are clustered according to weight and relationship, to create a concept cluster map" (Grech et al., 2002: 1719). Afterward, the concepts are sorted into themes. Eventually, the maps examined by the author and their statistical data analyzed the stability of the maps by re-running the algorithm several times and determining the best themes number (Wilden et al., 2018). On this foundation, content analysis was performed using BibExcel, Gephi, and Leximancer software, and the findings are presented in Section 3. After the mentioned analyses, the review is complemented with the interpretation of the bibliometric map.

Findings

The research results are presented in two parts below. The first is concept mapping, and the second is co-citation network analysis.

Concept Map

The network graph resulting in five different clusters is shown in Figure 1.1. In the following stage, a text-mining technique was applied using a Bayesian learning algorithm available in the Leximancer program (Wilden et al., 2017). First, the author evaluated the data presented by the program,

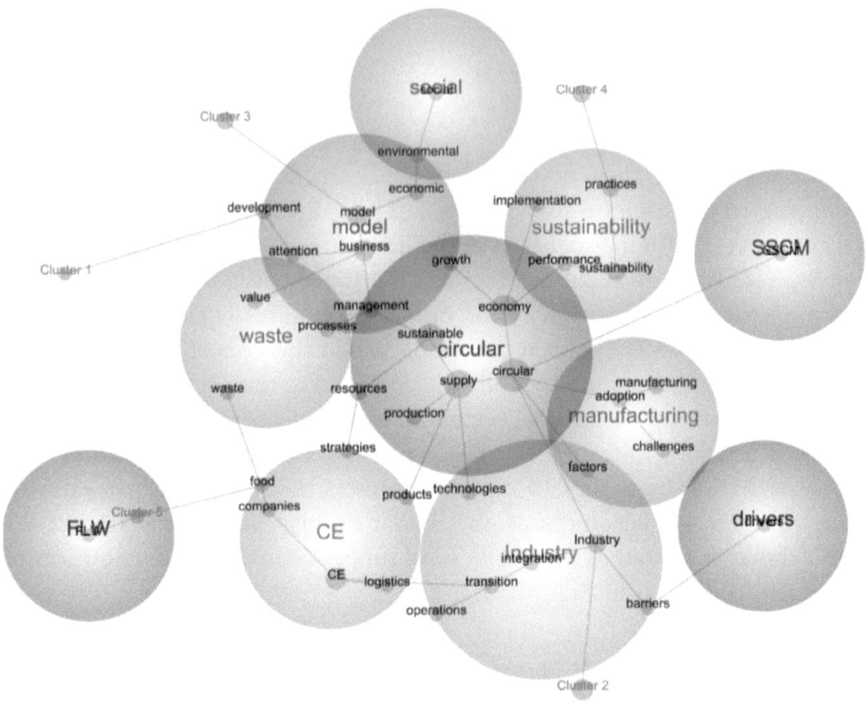

Figure 1.1. Concept Map
Source: Author's preparation

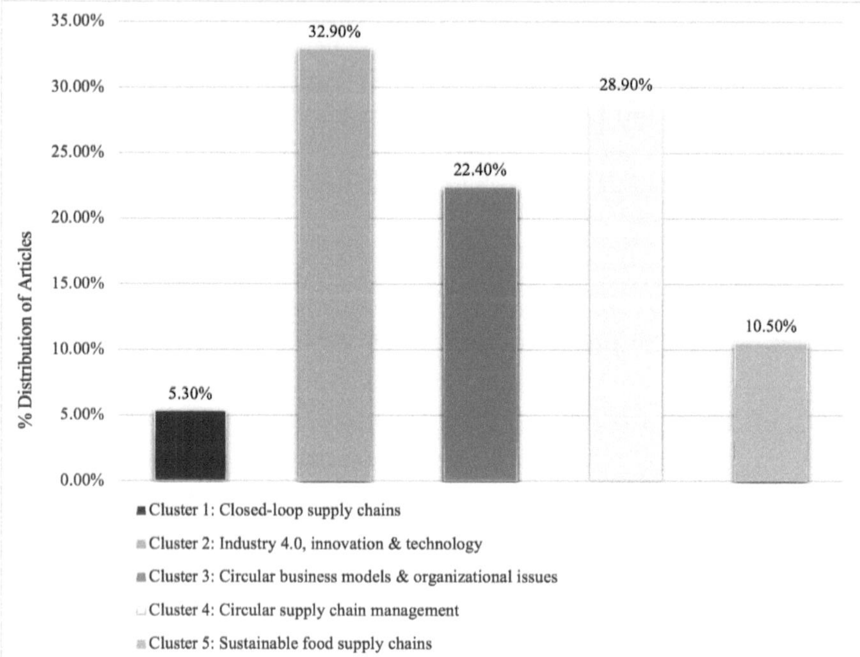

Figure 1.2. Theme analysis of the articles
Source: Authors preparation

and the texts in the clusters were read. Thus, the author named the clusters according to the topics they contained. Then, finally, to unveil the essential concepts and themes debated in the literature in each of the five clusters determined in the BC analysis (Figure 1.1). It presents a relatively unbiased and objective representation of the optimal structures within each cluster.

Analysis reveals five distinct clusters. When the concept map is examined, the clusters are named as follows, considering their sub-themes. Cluster 1 is represented by papers concentrating on closed-loop supply chains; cluster 2 includes industry 4.0, innovation and technology. Cluster 3 is primarily based on studies of circular business models and organizational issues. Cluster 4 is an emphasis on circular supply chain management. Finally, cluster 5 contains sustainable food supply chains. The distribution of the articles is shown in Figure 1.2.

Co-citation Network Analysis

Co-citation network analysis is defined as a type of explorative data analysis that depends on graph theory to investigate data generation (Pampel,

2004). A co-citation map comprises of nodes symbolizing papers and several edges delineating the cooccurrence of nodes in the bibliography of documents on that map (Leydesdorff, 2011). Documents are co-cited if they come together in the bibliography of others. It has been demonstrated that documents that are more frequently cited together by others are more probably to be related and therefore belong to a resembling content field (Ben-Daya et al., 2020; Fahimnia et al., 2015; Hjørland, 2013). Co-citation analysis of texts is extensively used to determine the most significant articles in each scientific area (Dong et al., 2015).

First, the data were imported from the Web of Science Core Collection database into BibExcel, designed to analyze bibliographic data (Persson et al., 2009). With BibExcel, one can complete a co-citation analysis by making pairs of the chosen items, such as commonly cited authors. Next, the references were retrieved from the source articles to analyze the literature co-citations. However, visualizing the whole network is not directly insightful because there are too many connections, making it difficult to see the underlying structure (Yu et al., 2014). Hence, to get a cleaner visualization for producing the graphs, the author filtered the articles considering the criteria of having at least seven common references and being linked with at least three articles to analyze effectively. As a result, 76 articles were obtained as the research sample. The final step was visualizing the co-citation structure as a research network based on the Gephi software. The location of each node and its color are used to classify the papers from the same cluster, while each node size relies on the weight of the papers (citations of each paper) (Maseda et al., 2022).

In Figure 1.3, each cluster is shown in different colors. For example, blue represents Cluster 2, while green represents Cluster 3. On the clusters indicated with varying colors in the diagram, the author and year information of the studies in the relevant field are also noted. In addition, the relations of these papers with each other in the context of co-citation are also depicted as a network structure.

It is evident that Cluster 2, which includes the themes of innovation, industry 4.0, and technology within the scope of supply chain and circular economy, has been the most prevalent subject that was discussed among the articles published in the WoS database. However, Cluster 1 related to closed-loop supply chains was the least discussed issue. When the prominent sub-topics within the clusters were examined, it was seen that the dimensions such as closed-loop supply chain management, knowledge base, knowledge dissemination, and remanufacturing were intensively addressed in the papers in Cluster 1.

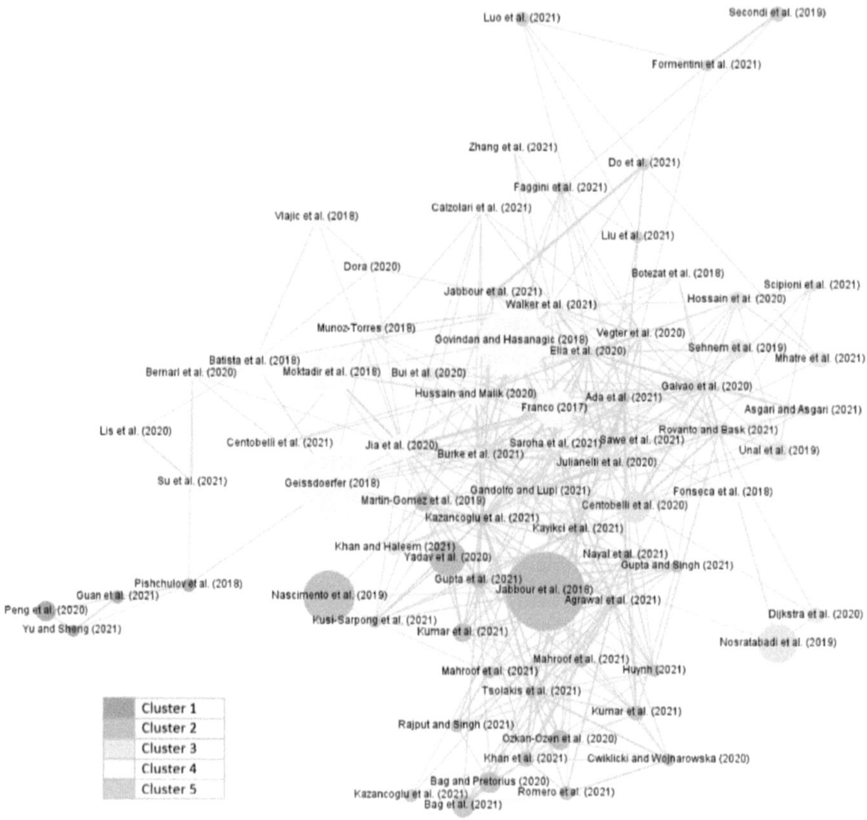

Figure 1.3. Visualized citation network
Source: Authors preparation

Industry 4.0 initiatives, models, and applications, big data, digital innovation, digital technologies, emerging technologies, technology integration, sustainable operations, decision making, drone, logistics, digital supply network design, CE 4.0, and manufacturing organizations, smart and sustainable CE, and clean production come to the fore among the topics discussed in the Cluster 2. The studies in Cluster 3 focused on the themes of business models, organizational learning, CE practices, adoption of CE, sustainable corporate finance, reverse logistics, operational excellence, value creation, small- and medium-sized enterprises, barriers, and incentives. Cluster 4 was the second largest group, discussed; supply chain management, product design, multinational enterprises, small- and medium-sized enterprises, business strategies, industrial ecology, circular interfirm networks, organizational enablers, collaboration, sustainable manufacturing practices, CE practices, global supply chain, green supply chain, integration, value recovery, and performance. Finally, the second least discussed theme among the articles published in the WoS database, Cluster 5, includes studies examining food loss and waste (FLW), FLW prevention and management, FLW supply and quantification, circular food supply chains, CE ecosystem, and the agri-food sector.

The size of each node is proportional to the number of citations received by the article. Thus, a large-sized node indicates a highly cited paper (Dong et al., 2015). According to the research findings, the articles with the highest impact value (Franco, 2017; Geissdoerfer et al., 2018; Govindan and Hasanagic, 2018; Lopes de Sousa Jabbour et al., 2018; Moktadir et al., 2018; Nascimento et al., 2019; Nosratabadi et al., 2019) are shown in Figure 1.3.

Conclusion

In their operational processes, businesses interact with all units that make up the supply chain, such as raw material providers, intermediaries, and customers. As a result, time, cost, and resource waste occur at every stage of the classical linear supply chain model (take, make, consume and dispose of) and it creates harmful effects on nature and the economy. Suppose the CE approach is applied in supply chains. In that case, it provides enormous resource savings, sustainable use of natural resources is ensured, costs are reduced, and living in a cleaner, and healthier environmental condition is possible. The CE requires an innovative transformation to supply chain structures to cope with the environmental threats caused by the traditional supply chain by addressing material circularity. The basic concept is to reduce and prolong material utilization while capturing and recreating novel values of goods and services throughout the supply chain (Khompatraporn, 2021).

The results of this study reveal; innovation, technology, and industry 4.0 play an essential role in the interaction of the supply chains and CE. For the implementation of 6R (recycle, remanufacture, reuse, recover, reduce, redesign) by transforming the supply chains into a closed-loop structure, there is a need to transform many activities such as waste management, material use reduction, new product development from recycled materials, packaging, transportation, reverse logistics, with an innovative perspective. Innovative technologies are fundamental for implementing more intelligent and collaborative logistics systems for materials and waste to recover their environmental and economic performances (Campos et al., 2020).

Innovative supply chains will be a significant research topic in the following years in both academia and industry. It has excellent potential to boost the productivity of manufacturing enterprises. The manufacturing industry practitioners are looking forward to innovative industry practitioners to boost their supply chain performance. The reorganization of current resources through business model innovation is vital to enabling a regenerative supply chain (Hofmann, 2019; Hussain and Malik, 2020).

The recognized gaps and future opportunities allow researchers, practitioners, and policymakers to expand research further. Research areas such as closed-loop supply chain and food waste management have been given much less attention in the literature. Therefore, these areas can provide abundant opportunities for research in future studies. Abreu and Ceglia (2018) propose that the CE requires adopting new business models and the significance for public authorities of providing sufficient support through environmental legislation, policies, and market and economic instruments. Partnerships among private and public institutions should be a path to reinforcing institutional capacities in building the CE. In addition, they suggest that establishing online platforms to increase knowledge sharing among partners and promoting knowledge sharing will enhance opportunities for industrial symbiosis between institutions. Because sharing knowledge among partners contributes to the co-development of new products by fostering an open innovation environment (Avunduk et al., 2021). The adoption of open innovation by partners in the supply chain, especially to discover and realize industrial symbiosis opportunities, is likely to contribute to the practitioners. This can contribute to the conversion of waste into value by using each other's wastes or by-products resulting from production by businesses working in different business lines. It is recommended for future studies to carry out qualitative and quantitative research to explore opportunities for industrial symbiosis by following an open innovation strategy among supply chain members. In addition, research on the development of innovative methods with the effective use of industry 4.0 tools in the supply chains can contribute to the literature.

One of the limitations of this paper is that merely the WoS database was used for collecting data. While WoS guarantees the quality of the articles, scanning other databases will enrich the amount of the data set. In that case, the results may indicate a more comprehensive understanding of the literature. The abstract, title, and references of the selected sample were analyzed with the program. The other limitation is that keywords and full text are not included in the analysis. For this reason, the inability to classify the methods used in the research is another limitation of the study.

References

Abreu, M. C. S. de, and Ceglia, D. (2018). On the implementation of a circular economy: The role of institutional capacity-building through industrial symbiosis. *Resources, Conservation and Recycling*, 138, 99–109. https://doi.org/10.1016/J.RESCONREC.2018.07.001.

Avunduk, Z. B., Kazan, H., Tatoglu, E., and Zaim, S. (2021). How Does External Knowledge Sourcing Affect Business Performance in Large-Scale Firms? The Mediating Role of Open Technology Strategies. *IEEE Transactions on Engineering Management*. https://doi.org/10.1109/TEM.2020.3047995.

Bastian, M., Heymann, S., and Jacomy, M. (2009). Gephi: An Open Source Software for Exploring and Manipulating Networks. *Proceedings of the International AAAI Conference on Web and Social Media, 3*(1), 361–62. https://ojs.aaai.org/index.php/ICWSM/article/view/13937.

Ben-Daya, M., Hassini, E., Bahroun, Z., and Banimfreg, B. H. (2020). The role of internet of things in food supply chain quality management: A review. 28(1), 17–40. https://doi.org/10.1080/10686967.2020.1838978.

Biesenthal, C., and Wilden, R. (2014). Multi-level project governance: Trends and opportunities. *International Journal of Project Management, 32*(8), 1291–1308. https://doi.org/10.1016/J.IJPROMAN.2014.06.005.

Blondel, V. D., Guillaume, J. L., Lambiotte, R., and Lefebvre, E. (2008). Fast unfolding of communities in large networks. *Journal of Statistical Mechanics: Theory and Experiment, 2008*(10), P10008. https://doi.org/10.1088/1742-5468/2008/10/P10008.

Boyack, K. W., and Klavans, R. (2010). Co-citation analysis, bibliographic coupling, and direct citation: Which citation approach represents the research front most accurately? *Journal of the American Society for Information Science and Technology, 61*(12), 2389–404. https://doi.org/10.1002/ASI.21419.

Dong, D., and Chen, M.-L. (2015). Publication trends and co-citation mapping of translation studies between 2000 and 2015. *Scientometrics*, 105, 1111–28. https://doi.org/10.1007/s11192-015-1769-1.

el Baz, J., and Iddik, S. (2022). Green supply chain management and organizational culture: a bibliometric analysis based on Scopus data (2001-2020*). International Journal of Organizational Analysis, 30*(1), 156–79. https://doi.org/10.1108/IJOA-07-2020-2307/FULL/PDF.

Ellen MacArthur Foundation. (2014). *Towards the Circular Economy Vol. 3: Accelerating the scale-up across global supply chains*. https://emf.thirdlight.com/link/t4gb0fs4knot-n8nz6f/@/preview/1?o.

Fahimnia, B., Sarkis, J., and Davarzani, H. (2015). Green supply chain management: A review and bibliometric analysis. *International Journal of Production Economics*, 162, 101–14. https://doi.org/10.1016/J.IJPE.2015.01.003.

Franco, M. A. (2017). Circular economy at the micro level: A dynamic view of incumbents' struggles and challenges in the textile industry. *Journal of Cleaner Production*, 168, 833–45. https://doi.org/10.1016/J.JCLEPRO.2017.09.056.

Gao, H., Ding, X. H., and Wu, S. (2020). Exploring the domain of open innovation: Bibliometric and content analyses. *Journal of Cleaner Production*, 275, 122580. https://doi.org/10.1016/J.JCLEPRO.2020.122580.

Geissdoerfer, M., Morioka, S. N., de Carvalho, M. M., and Evans, S. (2018). Business models and supply chains for the circular economy. *Journal of Cleaner Production*, 190, 712–21. https://doi.org/10.1016/J.JCLEPRO.2018.04.159.

Govindan, K., and Hasanagic, M. (2018). A systematic review on drivers, barriers, and practices towards circular economy: a supply chain perspective. 56(1–2), 278–311. https://doi.org/10.1080/00207543.2017.1402141.

Goyal, S., Chauhan, S., and Mishra, P. (2021). Circular economy research: A bibliometric analysis (2000–2019) and future research insights. *Journal of Cleaner Production*, 287, 125011. https://doi.org/10.1016/J.JCLEPRO.2020.125011.

Hjørland, B. (2013). Citation analysis: A social and dynamic approach to knowledge organization. *Information Processing and Management, 49*(6), 1313–25. https://doi.org/10.1016/J.IPM.2013.07.001.

Hofmann, F. (2019). Circular business models: Business approach as driver or obstructer of sustainability transitions? *Journal of Cleaner Production*, 224, 361–74. https://doi.org/10.1016/J.JCLEPRO.2019.03.115.

Hofmann, M., and Chisholm, A. (2016). *Text Mining and Visualization: Case Studies Using Open-Source Tools*. CRC Press Taylor&Francis Group.

Hussain, M., and Malik, M. (2020). Organizational enablers for circular economy in the context of sustainable supply chain management. *Journal of Cleaner Production, 256*, 120–375. https://doi.org/10.1016/J.JCLEPRO.2020.120375.

Kessler, M. M. (1963). Bibliographic coupling between scientific papers. *American Documentation, 14*(1), 10–25. https://doi.org/10.1002/ASI.5090140103.

Kraus, S., Breier, M., and Dasí-Rodríguez, S. (2020). The art of crafting a systematic literature review in entrepreneurship research. *International Entrepreneurship and Management Journal, 16*, 1023–42. https://doi.org/10.1007/s11365-020-00635-4.

Krippendorff, K. (2019). *Content Analysis: An Introduction to Its Methodology* (4th ed.). Sage Publications.

Kumar, B., Sharma, A., Vatavwala, S., and Kumar, P. (2020). Digital mediation in business-to-business marketing: A bibliometric analysis. *Industrial Marketing Management, 85*, 126–40. https://doi.org/10.1016/J.INDMARMAN.2019.10.002.

Lee, H. L., and Tang, C. S. (2017). Socially and Environmentally Responsible Value Chain Innovations: New Operations Management Research Opportunities. *64*(3), 983–96. https://doi.org/10.1287/MNSC.2016.2682.

Leydesdorff, L. (2011). Bibliometrics/citation networks. In *Encyclopedia of Social Networks*. SAGE Publications Inc.

Liu, Z., Yin, Y., Liu, W., and Dunford, M. (2015). Visualizing the intellectual structure and evolution of innovation systems research: a bibliometric analysis. *Scientometrics, 103*(1), 135–58. https://doi.org/10.1007/S11192-014-1517-Y/TABLES/5.

Lopes de Sousa Jabbour, A. B., Jabbour, C. J. C., Godinho Filho, M., and Roubaud, D. (2018). Industry 4.0 and the circular economy: a proposed research agenda and original roadmap for sustainable operations. *Annals of Operations Research, 270*(1–2), 273–86. https://doi.org/10.1007/S10479-018-2772-8/TABLES/3.

Luis, E. C., and Celma, D. (2020). Circular economy. A review and bibliometric analysis. *Sustainability (Switzerland), 12*(16). https://doi.org/10.3390/SU12166381.

Maditati, D. R., Munim, Z. H., Schramm, H. J., and Kummer, S. (2018). A review of green supply chain management: From bibliometric analysis to a conceptual framework and future research directions. *Resources, Conservation and Recycling, 139*, 150–62. https://doi.org/10.1016/J.RESCONREC.2018.08.004.

Maseda, A., Iturralde, T., Cooper, S., and Aparicio, G. (2022). Mapping women's involvement in family firms: A review based on bibliographic coupling analysis. *International Journal of Management Reviews, 24*(2), 279–305. https://doi.org/10.1111/IJMR.12278.

Masi, D., Day, S., and Godsell, J. (2017). Supply Chain Configurations in the Circular Economy: A Systematic Literature Review. *Sustainability 2017, Vol. 9, Page 1602, 9*(9), 1602. https://doi.org/10.3390/SU9091602.

Moktadir, M. A., Ali, S. M., Rajesh, R., and Paul, S. K. (2018). Modeling the interrelationships among barriers to sustainable supply chain management in leather industry. *Journal of Cleaner Production, 181*, 631–51. https://doi.org/10.1016/j.jclepro.2018.01.245.

Montecchi, M., Plangger, K., and West, D. C. (2021). Supply chain transparency: A bibliometric review and research agenda. *International Journal of Production Economics, 238*, 108–52. https://doi.org/10.1016/J.IJPE.2021.108152.

Murray, A., Skene, K., and Haynes, K. (2015). The Circular Economy: An Interdisciplinary Exploration of the Concept and Application in a Global Context. *Journal of Business Ethics 2015 140:3, 140*(3), 369–80. https://doi.org/10.1007/S10551-015-2693-2.

Nascimento, D. L. M., Alencastro, V., Quelhas, O. L. G., Caiado, R. G. G., Garza-Reyes, J. A., Lona, L. R., and Tortorella, G. (2019). Exploring Industry 4.0 technologies to enable circular economy practices in a manufacturing context: A business model proposal. *Journal of Manufacturing Technology Management, 30*(3), 607–27. https://doi.org/10.1108/JMTM-03-2018-0071/FULL/PDF.

Nosratabadi, S., Mosavi, A., Shamshirband, S., Kazimieras Zavadskas, E., Rakotonirainy, A., and Wing Chau, K. (2019). Sustainable business models: A review. *Mdpi.Com, 11*. https://doi.org/10.3390/su11061663.

Pampel, F. C. (2004). Exploratory data analysis. In M. S., Lewis-Beck, A., Bryman, and T. F. Liao (Eds.), *Encyclopedia of Social Science Research Methods*. SAGE Publications.

Persson, O., Danell, R., and Wiborg Schneider, J. (2009). How to use Bibexcel for various types of bibliometric analysis. In *Celebrating scholarly communication studies: A Festschrift for Olle Persson at his 60th Birthday* (pp. 9–24). ISSI. http://urn.kb.se/resolve?urn=urn:nbn:se:umu:diva-25636.

Randhawa, K., Wilden, R., and Hohberger, J. (2016). A Bibliometric Review of Open Innovation: Setting a Research Agenda. *Journal of Product Innovation Management, 33*(6), 750–72. https://doi.org/10.1111/JPIM.12312.

Rooney, D. (2005). Knowledge, economy, technology and society: The politics of discourse. *Telematics and Informatics, 22*(4), 405–22. https://doi.org/10.1016/J.TELE.2004.11.007.

Shang, T., Miao, X., and Abdul, W. (2019). A historical review and bibliometric analysis of disruptive innovation. *International Journal of Innovation Science, 11*(2), 208–26. https://doi.org/10.1108/IJIS-05-2018-0056/FULL/PDF.

Singh, S., Trivedi, B., Dasgupta, M. S., and Routroy, S. (2021). A bibliometric analysis of circular economy concept in E-waste research during the period 2008–2020. *Materials Today: Proceedings, 46*, 8519–24. https://doi.org/10.1016/J.MATPR.2021.03.525.

Smith, A. E., and Humphreys, M. S. (2006). Evaluation of unsupervised semantic mapping of natural language with Leximancer concept mapping. *Behavior Research Methods, 38*(2), 262–79. https://doi.org/10.3758/BF03192778.

Tsai, F. M., Bui, T. D., Tseng, M. L., Lim, M. K., and Hu, J. (2020). Municipal solid waste management in a circular economy: A data-driven bibliometric analysis. *Journal of Cleaner Production, 275*, 124–32. https://doi.org/10.1016/J.JCLEPRO.2020.124132.

Vogel, R., and Güttel, W. H. (2013). The Dynamic Capability View in Strategic Management: A Bibliometric Review. *International Journal of Management Reviews, 15*(4), 426–46. https://doi.org/10.1111/IJMR.12000.

Wilden, R., Akaka, M. A., Karpen, I. O., and Hohberger, J. (2017). The Evolution and Prospects of Service-Dominant Logic: An Investigation of Past, Present, and Future Research. *20*(4), 345–61. https://doi.org/10.1177/1094670517715121.

Wilden, R., Hohberger, J., Devinney, T. M., and Lavie, D. (2018). Revisiting James March (1991): Whither exploration and exploitation? *16*(3), 352–69. https://doi.org/10.1177/1476127018765031.

Yu, C., Davis, C., and Dijkema, G. P. J. (2014). Understanding the Evolution of Industrial Symbiosis Research. *Journal of Industrial Ecology, 18*(2), 280–93. https://doi.org/10.1111/JIEC.12073.

Zhang, N., and Zhao, Y. (2021). Green supply chain management in the platform economy: a bibliometric analysis. *Https://Doi.Org/10.1080/13675567.2021.1885635*, 1–17. https://doi.org/10.1080/13675567.2021.1885635.

Zhang, X., Chen, H., Wang, W., and Ordóñez De Pablos, P. (2016). *What is the role of IT in innovation? A bibliometric analysis of research development in IT innovation.* https://doi.org/10.1080/0144929X.2016.1212403.

Zhang, X., Yu, Y., and Zhang, N. (2020). Sustainable supply chain management under big data: a bibliometric analysis. *Journal of Enterprise Information Management, 34*(1), 427–45. https://doi.org/10.1108/JEIM-12-2019-0381/FULL/PDF.

Zhu, Q., Geng, Y., and Lai, K. hung. (2010). Circular economy practices among Chinese manufacturers varying in environmental-oriented supply chain cooperation and the performance implications. *Journal of Environmental Management, 91*(6), 1324–31. https://doi.org/10.1016/J.JENVMAN.2010.02.013.

Zhu, Q., Geng, Y., and Lai, K. hung. (2011). Environmental Supply Chain Cooperation and Its Effect on the Circular Economy Practice-Performance Relationship Among Chinese Manufacturers. *Journal of Industrial Ecology, 15*(3), 405–19. https://doi.org/10.1111/J.1530-9290.2011.00329.X.

Zupic, I., and Čater, T. (2014). Bibliometric Methods in Management and Organization: *18*(3), 429–72. https://doi.org/10.1177/1094428114562629.

Sustainable Consumer Behavior 2

YUSUF CAN ERDEM

Sustainability

There are several viewpoints on the definition and scope of sustainability, but a simple definition is: "Sustainability means meeting our needs without compromising the power of upcoming generations to meet their own needs for natural, social and economic resources." Since sustainability is not only environmentalism, the implicit concerns for social equity and economic development are also included in this definition (University of Alberta).

Today, social and environmental problems facing all of the societies across the world are so complicated that the term "sustainability" has been increasingly used to define the procedures and measures through which humanity avoids the exhaustion of natural resources, in order to be able to keep an ecological sense of balance. The overall aim is not to destroy the quality of life both today and tomorrow, but this is not possible only with improvements in areas like natural resources overexploitation. With a more comprehensive approach, modifications in any kind of manufacturing process and their implications on the use of energy and pollution, new consumption and purchasing patterns of consumers, the changing lifestyles, the attractiveness of investment decisions, the new business models and technological developments needed to enable the mentioned improvements are also included in the universal perception of sustainability (Sustainability—What is it? 2021).

This comprehensive concept of sustainability is actually based on three major pillars: the environment (planet), the economy (profit), and society (people). Both human impact on the environment and also its consequences on economic development were for the first time discussed in the UN Global meeting in Stockholm in 1972. In that critical meeting, it was declared: "Man is both creature and a molder of his environment, which gives him physical sustenance and affords him the opportunity for intellectual, moral, social and spiritual growth" (Declaration of the United Nations Conference, 1972).

One of the approved principles in the Stockholm meeting also stated that the natural resources of the earth must be protected for the welfare of both present and future generations and the governments were nominated as main authorities to assume the responsibility for environmental requirements.

The concern about the quality of life "tomorrow," which is a long-term oriented phrase is the critical difference between the concepts of "sustainability" and "sustainable development." The most comprehensive and a universally agreed upon definition of it is stated in the *Report of the World Commission on Environment and Development: Our Common Future* in 1987: "Sustainable development is the development that meets the needs of the present without compromising the ability of future generations to meet their own needs" (Our Common Future: Report of the World Commission on Environment, 1987).

Additional concepts like social justice, conservationism, and internationalism collaborated together by the end of the twentieth century to identify the more comprehensive notion of "sustainable development." In fact, this definition has two major components: The first one is the needs of the poor nations, which is a very urgent misery and the second one is the fact that social/cultural institutions and the state of technology in different countries enforce some limitations on the environment's ability to meet both present and future needs. It is very unfortunate that the basic needs of huge numbers of residents in the developing and underdeveloped regions are far from being satisfied. Furthermore, those people have justifiable aspirations for a better quality of life, which also asks for satisfaction by sustainable development.

By the end of the twentieth century, emerging markets like China and India have grown fundamentally with radically improved living standards while this growth put a pressure on the environment in general and on resources in particular. This overexploitation of natural resources essential for the achievement of economic growth and development has had a very negative impact on both the environment and the availability/cost of natural resources. This unusual improvement has made it clear that if consumption patterns are not more sustainable, the world resources and environment cannot keep up with this level of growth.

That is the topic of this chapter: sustainability in terms of consumer behavior.

Sustainable Customer

Thus, a deviation in both consumer behavior and lifestyle is essential to achieve sustainability and tackle climate change.

There are many different institutions and agents, including international organizations, governments, industries, trade unions, technologies, markets and others to achieve global sustainability. What is common for all those institutions is that their members and executives are all consumers at the same time with a critical role in the adoption of sustainable products and technologies. In fact, consumers across the world are both reasons and victims of the environmental issues of the industrialized world today due to the fact that more than half of the greenhouse gas emissions occur from household consumption. Moreover, households are also liable for considerable amounts of energy consumption and waste. In the European Union (EU), households consumed 25 percent of total energy in 2014 (Eurostat, 2016), while 8 percent of total waste in the EU in 2012 was also generated by households. (Eurostat, 2015) Consumers are also being targeted by huge amount of input regarding ethical and moral comments about life, consumption, and behavior in every stage of their lives. Therefore, customers in general deserve the privilege to be the major determinant of sustainability efforts (Antonides, 2017).

Being aware of this trendy issue, research companies have focused most of their research on sustainability. The concept of sustainability is so communal that the research firm Nielsen called 2018 "the year of the influential sustainable consumer" and foresaw that the world entering the decade of the sustainable shopper. According to the findings of a study from 2018 (Was 2018 the Year of the Influential Sustainable Consumer? 2018), 73 percent of global respondents said they were ready to change their habits of consumption just to reduce their environmental influence. This progressively spoken majority can transform to great opportunities for many brands that accomplish to effectively satisfy customer expectations with the sustainability of their products.

Almost half (48 percent) of U.S. respondents said they would either probably or definitely change their consumption routines to reduce their effect on the environment. According to the findings of a study in 2018 the total value of spending on sustainable fast-moving consumer goods (FMCG) market in the USA was $128.5 billion and this total value has been up by almost 20 percent with a compound average growth rate (CAGR). That is four times larger than conventional products (3.5 percent vs -1.0 percent comparatively) since 2014 thanks to such shoppers growing in both numbers and impact. By 2021, the Nielsen research team anticipated that the total spending on sustainable FMCG goods to reach $150 billion with an increase of $14 billion to $22 billion.

Researchers claimed that sustainable, organic, and clean attributes were the major features driving sales growth of sustainable product categories, which was almost 22 percent but was forecasted to reach 25 percent of store sales by 2021. Particularly, the sales share of sustainable products grew nearly 3 percent between 2014 and 2017 whereas the sales share of conventional products dropped by almost 4 percent (Was 2018 the Year of the Influential Sustainable Consumer? Nielsen, 2018).

Nielsen also conducted a different research to compare the 2017 and 2018 sales data of three different product categories—coffee, bath products and chocolate based on the sales value on the U.S. Food-Drugstore-Mass Merchandiser channels.

Figure 2.1 shows the sales growth rates of bath products in both value and sales units for 2017 and 2018. It was observed that the bath products with sustainability claims grew faster in terms of both criteria with respect to the category of total bath products.

The growth rate for whole bath products sales in value was 1 percent, while the sales of products with an essential oil presence statement grew at an impressive rate of 18 percent for the period of March 2017 to 2018.

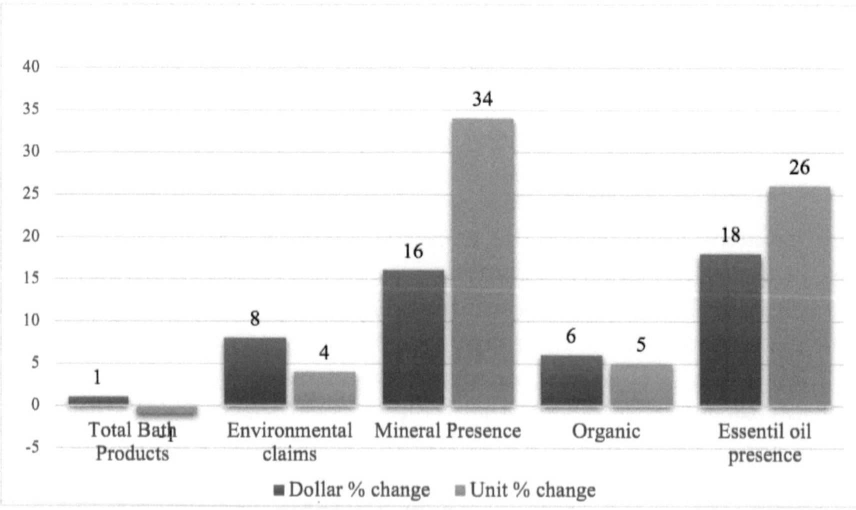

Figure 2.1. Sales Growth Rates: Bath Products 2017 vs 2018

Source: Nielsen answers on Demand (Health & Wellness characteristics). Total US Food-Drugstore-Mass merchandiser channels in 252-week period ending 24.3.2018 vs. year ago. (What's Sustainability Got to Do with It? Nielsen, 2018)

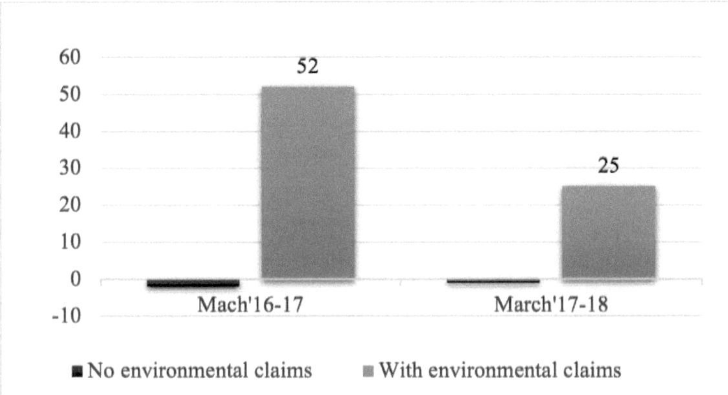

Figure 2.2. Growth in Dollar Sales—Coffee (%)

Source: Nielsen answers on Demand (Health & Wellness characteristics). Total US Food-Drugstore-Mass merchandiser channels in 52-week period ending 03.24.2018 vs. year ago. (What's Sustainability Got to Do with It? Nielsen, 2018)

Similarly, unit sales volume of those products grew 26 percent while the total category declined by 1 percent.

It can be concluded that including an essential oil in the product and claiming that additive on the product package caused the products to outperform competition based on the differences on both unit sales and the sales value of two groups of bath products.

A similar case was also observed in coffee consumption in the USA between March 2017 and March 2018 (Figure 2.2). The growth in dollar terms for the total coffee category for the mentioned period declined by 1 percent while coffee products with claims of fair trade and sustainable environment had a double-digit sales increase in value for the same period. Three of those sustainability claims for coffee brands were environmental claims, absence of non-natural ingredients and fair-minded trade. Overall, the growth rate of coffee products with claims of sustainability was higher than the overall growth rate of the coffee product category.

According to online research conducted by Coleman Parks on behalf of Accenture in 2019 with a sample of 1,500 U.S. consumers. Consumers are both more conscious about the impact of their purchases on environment and more willing to make sacrifices with respect to price and opportunism for the sake of eco-consciousness. Some of the major findings of this research are below:

The American respondents expressed their preference to purchase at retailers who value sustainability and this trend seems to be lasting. What is meant by "sustainability" is "environmental friendliness" and the retailers were also expected to show how environmentally friendly they were by almost half of the respondents (47 percent).

Similarly, 54 percent of the respondents said they welcomed in-store recycling services while 54 percent of them expressed their intention to use such services.

Same-day shipping services offered by some global retailers were appreciated by consumers at first but half of them declared that they would have sacrificed such speedy services and preferred regular shipping or in-store pickup just to benefit the environment (Chochrek, 2019).

Sapio Research conducted online research in the beginning of 2019 with a sample of 1346 business executives from six different countries in Europe, America and Australia and 6218 internet users aged eighteen and over.

According to the findings of a survey almost half (47 percent) of internet users in the United States, the United Kingdom, Australia, France, Italy, Spain, Mexico, and Germany stated that they had changed their consumption preferences since companies violated their personal values (Figure 2.3). Among the reasons for the switch, the top one was "Protect-

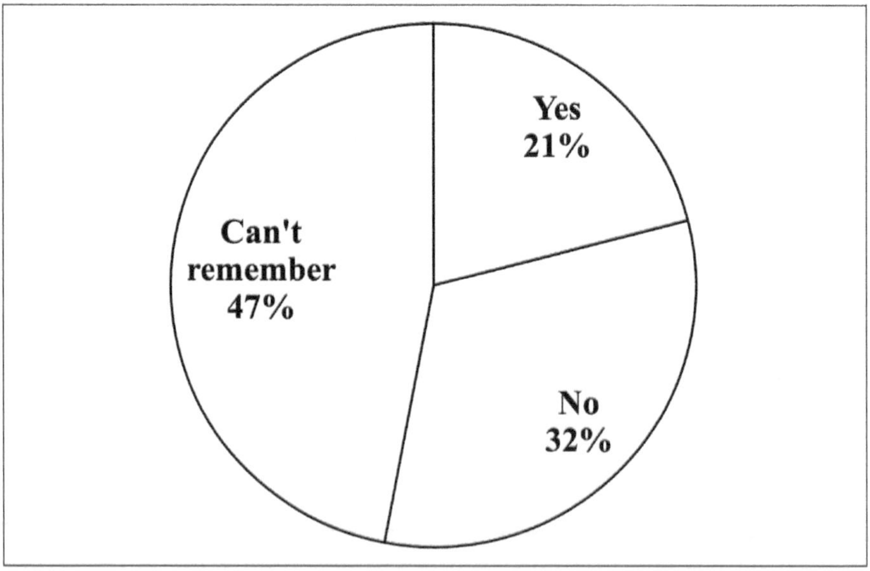

Figure 2.3. The number of Internet users who have switched a product/service because the company violated their personal values (January 2019).
Source: www.emarketer.com/chart/228374/internet-users-worldwide-who-have-switched-productservice-company-violated-their-personal-values-jan-2019-of-respondents

Table 2.1. Reasons to switch product/service

Reasons	%
Protecting the environment	11
Lack of transparency	9
Climate change	5
Corporate social responsibility	5
Data security	5
Ethnic discrimination & equality issues	5
Financial corruption	5
LGBTQ discrimination	5
Toxic workplace culture	5

Source: www.emarketer.com/chart/228374/internet-users-worldwide-who-have-switched-productservice-company-violated-their-personal-values-jan-2019-of-respondents

ing the environment," and "the concerns about climate change" was cited by only 5 percent of the respondents (Table 2.1).

Since the concept of "sustainability" is an all-inclusive one, it is essential to define the sustainable attributes since they reflect the changing priorities of consumers and different aspects of the overall and ever-expanding concept.

In April 2019, Nielsen published a study on more than 21,000 U.S. households, following the analysis of consumers' attitudes toward sustainable attributes across thirteen product categories. Two of the top sustainable attributes, "being environmentally friendly" and "recycled packaging" were stated as the most prior attributes by the respondents who expressed their willingness to make an additional payment if needed. However, in each group, this precedence was different from the general trend.

For instance, in some food categories like "eggs" consumers prioritized "farming aspects" as more important compared to "technological sustainability aspects." While "environmentally friendly" was the top ranked attribute across all the Consumer-Packaged Goods (CPG) categories, consumers listed "farm raised" and "organic farming" as the top two attributes when it came to purchasing eggs. "Being environmentally friendly" was not in the top four attributes in the eggs category. This indicates that the priorities of consumers on sustainability aspects may range from the ones with a straighter influence on consumers' personal health to more macro ones with global effects.

In the meantime, the ranking in a household goods group like laundry was different since respondents prioritized world well-being over their personal health. Consistent with the CPG overall, "environmentally friendly" and "recycled packaging" were ranked as the most significant features consumers were willing to spend on in laundry. This is a confirmation of the finding that priorities of consumers do change based on the perception of the sustainability attribute.

The Generational Differences among Customers

The term "consumer" is a very broad one even from a global perspective. Any research regarding the consumption forms of consumers in general is supposed to make a segmentation analysis in terms of either demographic and/or psychographic bases. Given the fact that demographic configuration of the world population keeps changing at a very high speed, it is crucial to search for the differences between consumers in different generations as far as their sustainable purchase intentions are concerned.

The findings of the research by Nielsen in 2018 revealed that the number of millennial respondents who claimed that they were changing their consumption patterns to decrease the effect on the environment were almost twice as the number of baby-boomer respondents (75 percent vs. 34 percent). Millennials also expressed their will to make additional payments for products containing sustainable or environmentally friendly ingredients (90 percent vs. 61 percent), natural or organic ingredients (86 percent vs. 59 percent), or products with social responsibility claims (80 percent vs. 48 percent) compared to the baby-boomers. The third major criteria to compare two generations was about how different they were regarding their willingness to switch their favorite brands to buy environmentally friendly products. As expected, millennials not only were more willing to change their brands for the sake of buying environmentally friendly products (53 percent vs. 34 percent) but it was also easier for them to find those products in the stores compared to baby-boomers (74 percent vs. 46 percent).

This last finding in particular is an indicator that the split of generations in sustainable consumption may be driven by technology (Figure 2.4). The mentioned research showed that the majority (67 percent) of the sustainable shoppers/respondents were more likely digitally involved, which meant that they were capable of easily having both the products and the knowledge needed thanks to their experience and abilities in online shopping and data management (Was 2018 the Year of the Influential Sustainable Consumer? 2018).

According to the findings of the research by NielsenIQ and Natural Marketing Institute (NMI), 60 percent of Americans are considered in the "Sustainable Mainstream" category. Although they wish to be more sus-

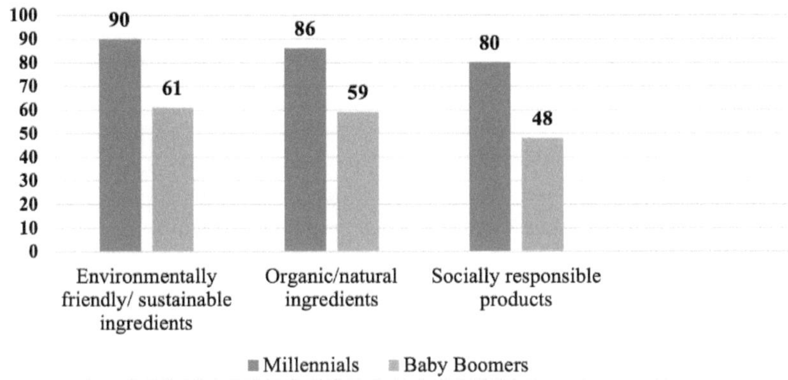

Figure 2.4. Willingness to pay more for:
Source: Nielsen: Was 2018 the Year of the Sustainable Consumer?

tainable, they are still looking for some additional benefits, like improved health or environmental or cost savings.

The differences among consumers from different generations have been popular research topics for researchers.

The research company Kearney conducted a study with 1,500 Canadian and U.S. consumers in 2019 regarding the preferences of Gen Z consumers for wellness and retail. Some of these findings are below:

- Sixty-seven percent of respondents expressed their preference to purchase products made with ingredients they can understand.
- Regarding the packaging, 58 percent of the Gen Z respondents preferred eco-friendly packaging while 65 percent wanted that packaging to be simple.
- Environmentally sustainable products were said to be searched by 57 percent of the respondents and almost half of them said they wanted locally sourced or made products.
- An interesting response of the Gen Z respondents was that although they were interested in eco-friendly, locally sourced, and sustainably packaged products, only 38 percent of them strongly agreed/agreed with the statement "I am willing to pay a premium for locally-sourced products." The percentage of the respondents ready to pay a premium price for sustainably packaged products was even lower (33 percent).

The More Recent Data

The global research company Deloitte conducted similar surveys to find out the key findings and the changes in sustainability and consumer behavior in March 2020 and March 2021 with the following findings:

- Sustainability is still a vital concern for consumers in 2021 as it was in 2020. Thirty-two percent of consumers in 2021 said they were very much involved with embracing a more sustainable way of life.
- In 2021, 28 percent of the respondents said they stopped purchasing certain products just because of some moral or environmental concerns.
- In 2020, ethics was a major issue more to younger consumers. The respondents at the age group of eighteen to thirty-four years were most likely to choose brands based on their ethical values, and decrease their animal-product consumption. Similarly, in 2021 Gen Z adopted more sustainable behaviors compared to the other age groups: 50 percent reduced their purchases and 45 percent quit buying certain brands because of sustainability or ethical concerns.
- Avoidance of single-use plastics is still the most common way shoppers display their assurance to sustainability; 61 percent in 2021 and 66 percent in 2020. Almost half of the respondents expressed their motivation on buying local goods (45 percent) and seasonality (49 percent) in 2021.
- In general, lack of interest is still the key barrier to adopting a sustainable lifestyle with an increase in 2021. In 2020, 16 percent of the respondents said "they were not interested" and the rate was 22 percent in 2021.The following barriers are "perceived expense" (16 percent) and "issues accessing relevant information" (15 percent). Regarding the switch to renewable energy and buying locally produced food, high prices and no availability were listed as the main barriers. (Shifting sands: Are consumers still embracing sustainability 2021?)

It is also clear that consumer interest in sustainability is not only limited to products or services but it is also considered essential for anything from sourcing for suppliers to distribution modes and even accessories. One of the interesting examples is the trend of #Gostrawless. Knowing that the daily use of plastic straws only in the United States is long enough to circle the Earth 2.5 times and it is one of the top 10 items identified during beach cleanups across the World, refusing the use of plastic straws is a very simple but meaningful way of caring for sustainability and that is what happened in many cities in the United States (Fighting for Trash-free Seas).

Therefore, no matter how it is defined, sustainability is top of mind for consumers around the globe, according to research. Therefore, these should be taken as significant messages by businesses since the shift in both consumer intentions and purchasing trends of consumers in favor of sustainability already have major impacts in shaping business settings. No matter the industry or the size, all of the profit/nonprofit oriented organizations either affecting or to be affected by this wave need to adapt themselves immediately. Both practitioners and researchers have been monitoring this process of adaptation from different perspectives (https://nielseniq.com/).

Sustainability as a Module in a Customer Experience Journey

From the perspective of marketing theory, it can be said that the trend of sustainable consumption can be considered as a module of the Experience Marketing Era, which is grounded in the fact that customers form their satisfaction and loyalty on every stage of the experience they carry on not only with the brand but also with the overall performance of the company. What is very critical to a product or service's feasibility is how the customer feels about the product/service, starting with the very first exposure and throughout the all-inclusive route of purchasing, consumption and afterwards (Tavsan, Erdem, 2018). The perceived sustainable image of any institution may be a very major stage in that process, which is also called the "customer journey." Executives in charge of many departments with or without direct customer touch are responsible to map and shift every customer-touch point to a spot of pleasure so that the whole customer experience may be an entertaining journey. This is an essential step on the way to loyalty and the image of sustainability may be a great advantage to provide a great customer experience at least for some customer segments. In other words, while the concept of marketing in both theory and practice is in a transition from "what it is" to "how it is," companies may enjoy the contribution of sustainability to survive and prosper in today's cutthroat competition.

Researchers have identified some steps to carve out the sustainability paths and stay on them in order to embrace sustainability as a vital module in customer experience management practice.

The first step brings up the similarity of "sustainability" with "Customer Experience Management (CXM)" in a very solid way: both need holistic approaches since neither of them can be succeeded with the efforts of only one department. Besides, both ought to be included in both job descriptions and performance evaluation processes of employees across the company. In other words, every member of the organization should feel the determination to consider sustainability at every stage ranging from planning to disposal phases. Since all direct and indirect experiences of the customers accumulate to form a sustainable image, all the activities and performance of employees are considered as parts of a big picture that will be interpreted from various customer perspectives. What makes this task more difficult is the fact that all the members of the value chain including different tiers of suppliers, distributors, wholesalers and retailers including marketing intermediaries constitute the parts of this big picture. What is needed is to put a sustainability lens on all the major decisions like the design of the products and shipment alternatives. That sustainability lens should also be used to monitor all the business processes and the product/service providers with a direct/indirect impact on the supply chain. This is an obligation to be carried out by the logistics management that is a discipline of supply chain management with an aim to plan, organize, implement, and control the flow of resources (raw materials, semi-finished and finished products, services and associated information) from the point of origin to the point of consumption with a mission to satisfy the requirements of customers (El-Berishy et al., 2013). In fact, this holistic approach is both critical and demanding since it needs a seamless coordination of various entities with conflicting operating models and business goals. Raw material suppliers, transportation companies, wholesalers, manufactures, service providers, retailers and even the cargo companies are all expected to operate as one entity in order to achieve the objective of sustainability. The only way to scrutinize the competencies of all the network members and to fine-tune their investments in order to create new inter-company partnerships focused on social responsibility and environmental impact is to adopt "Green Logistics."

The concept of green logistics has been around since the 1990s and continues to grow in importance as businesses become more aware of how their activities impact the environment. According to Srivastava, Green Supply Chain Management (GSCM) is:

> Integrating environmental thinking into supply-chain management, including product design, material sourcing and selection, manufacturing

processes, delivery of the finished product to the final users as well as the management of disposal stage. (Srivastava 2007, p. 54)

Greening the supply chain is a main challenge for managers of many profit and/or nonprofit institutions today. The main concern is how to escalate an organizational environmental consciousness and to integrate environmental management practices into their logistics activities (Zhu et al., 2008 a, b; Zhu, Lai, 2008). The good news is that many companies across the globe are increasingly recognizing the fact that environmental management is a fundamental issue with the potential for a long-term impact on organizational performance. For example, the environmental management system standard of ISO14001 has been implemented by more than 40,000 companies across the globe (Zhu and Sarkis, 2006). ISO 14001:2015 can be used by any organization trying to manage the environmental duties in an organized way that contributes to the environmental pillar of sustainability with the intended outcomes regardless of its industry, size and nature. The planned objectives can be listed as below:

- improvement of environmental performance,
- satisfaction of obedience requirements,
- attainment of environmental goals.

Some popular green supply chain management applications also include decreasing waste, supplier valuation on the base of environmental performance, development of eco-friendly products, and reduction in carbon emissions of transportation vehicles (Walker et al., 2008).

Green Supply Chain Management may be very effective and efficient if and only if it is implemented in coordination throughout the supply chain in a consistent and a holistic way (Davies and Hochman, 2007), which may be encouraged by both national and regional legislators. A good example is what happened in the European Union. In 1993, the EU published "European Eco-Management and Audit Scheme-EMAS," the most trustworthy and strong environmental management tool which includes some elements in addition to the official requirements of the international standard ISO 14001. The very strict requirements of EMAS include:

- Legal obedience with all environmental legislation, approved by an official authority
- Nonstop progress of environmental performance

- Confirmation of the performance by a definite competent verifier
- Publication of basic environmental data in a yearly report

As of June 2021, a total of 3.851 organizations and 12.856 sites have been registered in the EMAS scheme with a growth rate of 0.3 percent for organizations and 0.8 percent for sites since September 2020 (European Commission: Statistics and Graphs, 2021).

Even more, "sustainability" is not a static concept since it is never checked off as forever. It rather needs to be updated continuously because customers and the market are constantly changing. For example, it is not sufficient any more to be "healthy for me" since what is expected is to be "healthy for us" now.

The second step, according to Nielsen researchers, is the identification of sustainability for the brand. Having said that sustainability embraces many interconnected facets across being environmentally conscious, having recyclable packaging and many more, companies need to recognize the right one with the best possible impact on the way to be sustainable. This right choice will be based on the industry, the company positioning strategy, the product portfolio and the targeted customer profile. Being a service-oriented company, Starbucks Coffee Company has officially abandoned single-use plastic straws from all its stores by designing a straw-less lid or straw options made of different material. In this way, Starbucks took the option of literally touching their customers in a sustainable way. In the food and beverage industry, waste food can be turned into work-in-process and semi-finished goods like animal feed, protein flour, or even beer (Haller, Lee, and Cheung, 2020). A tire manufacturer could prefer to invest in growing rubber trees or in chemical research to minimize the pollution in manufacturing processes.

The identified sustainability feature should better be put into life in every touch point with both internal and external customers starting from production to the disposal stages in factories, warehouses, transportation, retailers and even in kitchens/refrigerators. With the digital revolution, websites, and even cargo companies are also included in this life cycle.

Retailers already started using the latest technologies to provide traceability of source materials and transparency of production methods so that customers may learn about the origins of their favorite foods. Online shoppers may be offered the option of "the last-mile delivery" to provide them trade-offs between supply chain costs, service, and environmental impact. Therefore, the customer may enjoy opportunities to participate in sustainability efforts by reducing some emission and pollution.

Leading manufacturers are discovering new ways to achieve measurable growth without a larger footprint by using state-of-the-art digital tools with flexible production systems leading to eco-efficiency through resource optimization (The manufacturers lighting a path to sustainable growth, 2021).

The emissions-reduction efforts and other green-specific measures are also considered as regular controls leading to cleaner production (Widlitz, 2020; Haller, Lee, Cheung, 2020).

The final step on the way to accomplish integrated sustainability across the company and the supply chain will be to keep testing all the links and taking necessary measures to update the end-to-end process. It is similar to monitoring and modifying customer journey maps, a major step in Customer Experience Management with an exception that the network to be monitored for sustainability is much larger depending on the structure of the supply chain. Thanks to consumer insights, the necessary adjustments to build a sustainability strategy into the brand messaging and portfolio optimization will be carried on so that the brands in particular and the company in general may resonate genuinely across all the consumers touch points. In other words, the consumers' sustainability expectations and their perception of the direct/indirect experiences will be redesigning sustainability development (Wilson, 2019).

Conclusion

It is globally true that companies in various industries have to deal with both the never-ending expectations of their customers for convenience, and also increasing/changing demands from governments, international/local institutions, shareholders and consumers to address environmental issues. Taking necessary measures and following up the implementation of them in both local and global scales are also the basic requirements to be fulfilled and this is a never-ending task with no known deadline. Besides, this requirement for sustainability is already a major advantage in the battle for the ever-shrinking and differentiated demands from customers. As the findings of research show, not only the marketers but all the business executives better be aware of this new battle ground with new rules of the war-game since consumers care about their overall satisfaction based on every single stage of experience with the brands. Integrating "sustainability" into the satisfaction bundle provided by companies seems to be the major contribution on the way to reach a sustainable world which may already be a dream too late to come true.

References

Antonides, G. (2017). Sustainable Consumer Behavior: A Collection of Empirical Studies. Sustainability, MDPI Books, November. Retrieved from https://doi.org/10.3390/books978-3-03842-584-7.

Betti, Francisco, Enno de Boer, and Yves Giraud. "The manufacturers lighting a path to sustainable growth," last modified March 15, 2021. Retrieved from: https://www.mckinsey.com/business-functions/operations/our-insights/the-manufacturers-lighting-a-path-to-sustainable-growth.

Chochrek, E. (2019). Why Sustainability Should Be Top of Mind for Retailers This Holiday Season. Last modified October 1, 2019. Retrieved from: https://footwearnews.com/2019/business/retail/holiday-shopping-2019-sustainable-fashion-retail-1202850297/.

"The Database: The Business of Sustainability," last modified May 6, 2019. Retrieved from: https://nielseniq.com/global/en/insights/multimedia/2019/the-database-the-business-of-sustainability/.

Davies, J., and Hochman, S. (2007). The greening of the supply chain. Supply Chain Management Review, 11(5), p.13–14.

"The Declaration of the United Nations Conference on Human Environment," last modified June 16, 1972. Retrieved from: https://legal.un.org/avl/ha/dunche/dunche.html.

El-Berishy, N., Rügge, I., Bernd Scholz-Reiter, B. (2013). The Interrelation between Sustainability and Green Logistics, IFAC Proceedings Volumes, 46, 24, pp. 2013, 527–31, ISSN 1474-6670, ISBN 9783902823502, https://doi.org/10.3182/20130911-3-BR-3021.00067.

European Commission: Statistics and Graphs. (2021). Last modified June 30, 2022. Retrieved from: https://ec.europa.eu/environment/emas/join_emas/faqs_en.htm#s2q1.

European Commission: Statistics and Graphs. (2021). Last modified June 30, 2022. Retrieved from: https://ec.europa.eu/environment/emas/emas_registrations/statistics_graphs_en.htm.

European Commission: Statistics and Graphs. (2021). Last modified June 30, 2022. Retrieved from: https://www.emas.de/fileadmin/user_upload/4-pub/UGA_Infosheet_From-ISO-14001-to-EMAS.pdf.

"Fighting for Trash-free Seas," Ocean Conservancy, June 30, 2022. Retrieved from: https://oceanconservancy.org/trash-free-seas/.

Haller, K. Lee, J., and Cheung, J. (2020). ""Meet the 2020 consumers driving change," IBM Institute for Business Value. Retrieved from: https://nrf.com/research/meet-2020-consumers-driving-change?utm_medium=Twitter%20NRF%20Organic&utm_source=Social&utm_campaign=NRF%20Research&utm_content=IBM%20Meet%202020% 20Consumer.

"Internet Users Worldwide Who Have Switched a Product/Service Because the Company Violated Their Personal Values, Jan 2019 (% of respondents)," last modified March 19, 2019. Retrieved from: https://www.emarketer.com/chart/228374/internet-users-worldwide-who-have-switched-productservice-company-violated-their-personal-values-jan-2019-of-respondents.

"Our Common Future: The Report of the World Commission on Environment," last modified 1987. Retrieved from: https://sswm.info/sites/default/files/reference_attachments/UN%20WCED%201987% 20Brundtland%20Report.pdf.

"Retailers Get Serious About Sustainability Into 2020," last modified February 3, 2020. Retrieved from: https://www.forbes.com/sites/staceywidlitz/2020/02/03/retailers-get-serious-about-sustainability-into-2020/?sh=78c7a9ac5e09.

Shifting sands: Are consumers still embracing sustainability 2021? (2021) Deloitte. Retrieved from: https://www2.deloitte.com/uk/en/pages/consumer-business/articles/sustainable-consumer.html.

Srivastiva, S. (2007). Green supply-chain management: A state-of-the-art literature review. International Journal of Management Reviews, 9: 53-80. https://doi.org/10.1111/j.1468-2370.2007.00202.x.

"Sustainability—What is it? Definition, Principles and Examples," last modified June 18, 2021. Retrieved from: https://youmatter.world/en/definition/definitions-sustainability-definition-examples-principles/.

Tavsan, N., Erdem, C. (2018). Customer Experience Management: How to Design, Integrate, Measure and Lead. Tasora Books, MN, USA. ISBN-13: 978-1934690956.

Walker, H., Di Sisto, L., McBain, D. (2008). Drivers and barriers to environmental supply chain management practices: Lessons from the public and private sectors, Journal of Purchasing and Supply Management, 14, (1), p.69-85. ISSN 1478-4092, https://doi.org/10.1016/j.pursup.2008.01.007.

"Was 2018 the Year of the Influential Sustainable Consumer?" last modified December 17, 2018. Retrieved from: https://nielseniq.com/global/en/insights/analysis/2018/was-2018-the-year-of-the-influential-sustainable-consumer.

"What's Sustainability Got to Do with It?" last modified October 16, 2018. Retrieved from: https://nielseniq.com/global/en/insights/report/2018/whats-sustainability-got-to-do-with-it/.

"What is Sustainability?" University of Alberta, Office of Sustainability. Accessed June 30, 2022. Retrieved from: https://www.mcgill.ca/sustainability/files/sustainability/what-is-sustainability.pdf.

Widlitz, S. "Retailers Get Serious About Sustainability into 2020," last modified February 3, 2020. Retrieved from: https://www.forbes.com/sites/staceywidlitz/2020/02/03/retailers-get-serious-about-sustainability-into-2020/?sh=3b6c4f6f5e09.

Wilson, Julia "A Sustainability How-To Guide for Retailers and Manufacturers," last modified April 23, 2019. Retrieved from: https://www.triplepundit.com/story/2019/sustainability-how-guide-retailers-and-manufacturers/83256/.

Wilson, J. "A Sustainability How-To Guide for Retailers and Manufacturers," last modified April 24, 2019. Retrieved from: https://nielseniq.com/global/en/insights/analysis/2019/a-sustainability-how-to-guide-for-retailers-and-manufacturers/.

Zhu, Q., Sarkis, J. and Lai, K. (2008). Confirmation of a Measurement Model for Green Supply Chain Management Practices Implementation. International Journal of Production Economics, *111*, p. 261-273. https://doi.org/10.1016/j.ijpe.2006.11.029.

Zhu, Q., Sarkis, J., Cordeiro, JJ., Lai, K. (2008). Firm-level correlates of emergent green supply chain management practices in the Chinese context. Omega, *36*(4), p. 577–91.

Zhu, Q., Sarkis, J., Lai, K. (2008). Green Supply Chain Management Implications for closing the Loop. Transportation Research Part E: Logistics and Transportation Review, *44*(1), p. 1–18.

3

An Extensive Review of the Applications of Circular Economy Concepts in Turkey

ALPER ALTINANAHTAR

Introduction

Expanding markets increase the need for new, finite resources and require advancement of new technologies, increase the cost of production, and make it harder to survive as a producer and as a consumer. Advancements in extraction of resources and production technologies are bound to fail if we run out of resources. Prevailing linear economy model, "take-make-dispose (waste)," is failing because finding water, food, and especially fossil fuel sources is getting harder. Biological diversity is diminishing, and financial systems can no longer provide an alternative for them. Thus, alternative strategies should be implemented to offset the negative effects of market expansions and depletion of resources. Thus, implementation of circular economy concepts becomes inevitable. Kirchherr, Reike, and Hekkert (2017) have identified 114 different definitions of circular economy and most of these definitions revolve around the 3 R's; reduce, reuse, recycle. The Ellen Macarthur Foundation (2017) used similar three principles to define a circular economy: "Design out waste and pollution, keep products and materials in use, regenerate natural systems." The European Commission, on the other hand, used the 9R Framework (Refuse, Rethink, Reduce, Re-use, Repair, Refurbish, Remanufacture, Repurpose, Recycle, and Recover), which was initially developed by Potting, Hekkert, and Worrell (2017) to categorize circular economy strategies.

According to the Ellen Macarthur Foundation (2017), transition to circular economy will have economic, environmental, and social benefits. For instance, it will induce economic growth, increase material cost savings, result in the creation of more jobs, increase land productivity, yield better soil health, reduce pollution, increase profits, reduce volatility and improve customer perception/loyalty and result in more demand for business services. This transition is estimated to generate up to 630 billion USD from the net-material cost savings in the sectors of complex medium-sized products in the EU and up to 700 billion USD from potential savings in fast moving consumer goods globally.

Section 2 will provide information in regards to some of the important dates and organizations in Turkey's transition to Circular Economy. In section 3, some of Turkey's major groups and their companies' CE implications will be reported. Section 4 will provide information about the general assessment of Turkey's transition to circular economy and conclude.

Circular Economy Chronology in Turkey

The largest greenhouse gas (GHG) emitter in the world, China—followed by many other economies with very high carbon dioxide (CO_2) emissions—has led 196 countries to create strategies with low carbon development (K. Winans, A. Kendall, and H. Deng, 2017). Accordingly, since 1980s, Turkey has been working on establishing an institutionalized environmental authority, which would allow the country's implementation of circular economy concepts (K. Winans, A. Kendall, and H. Deng, 2017). Waste recycling and recovery, renewable energy utilization, battle against climate change and many other targets have been identified to support CE implementation. After the adoption of the Circular Economy Action Plan by the European Union in 2015, in 2016, in Turkey the Circular Economy concept was first introduced by the Business Council for Sustainable Development Turkey (BCSD Turkey) and the European Bank for Reconstruction and Development (EBRD) (Güngör, 2019). The Zero Waste Project was initiated by the Ministry of Environment and Urbanization in 2017. It was one of the largest implementations of CE concepts in Turkey (Env.net, 2020). By 2019, 18,750 public institutions had joined the project. The project mainly focused on the prevention of materials waste and if waste is inevitable then it should be kept at the minimum and it should immediately be separated and recycled at its source to be used as raw material or energy[1] (Nuroğlu, 2019).

In 2018, Business Council for Sustainable Development Turkey (BCSD Turkey) and Holland Innovation Network Turkey CIRCI organized a Circular Design Workshop, which was provided by CIRCO where representatives of the twenty-biggest Turkish companies shared their thoughts about circularity (Güngör, 2019). Around the same period, Türkiye Materials Marketplace (TMM), an online platform that is managed by BCSD Turkey with the funding support of the EBRD, came online. The goal of the platform is to make one organization's waste and/or by-product, another organization's raw material through material exchanges. In 2020, Business for Goals Platform and DCUBE together published "Circular Economy Manual for Businesses." The aim of the manual is to accelerate the transition to the circular economy in Turkey and to ensure that the concept of circular economy is popularized both in the business world and in public (Hedefler İçin İş Dünyası, 2020). In 2019, The European Commission adopted a set of proposals, "A European Green Deal," to rearrange several EU polices to achieve a net greenhouse gas emissions by at least 55 percent by 2030, compared to 1990 levels where the main aim is to become the first climate-neutral continent by 2050. Lowering green gas emissions and growth strategies are listed under seven policies:

1. Clean energy
2. Sustainable industry
3. Building and renovation
4. From field to table
5. Eliminating pollution
6. Sustainable movement
7. Biological diversity (European Commission, 2019)

In 2020, under the coordination of the Ministry of Commerce with the participation of several ministries a workshop was organized and The Green Deal Action Plan, "the first major publication regarding harmonization with the European Green Deal and acknowledging the importance of circular principles"[2] was published in 2021 (T.C. Ticaret Bakanlığı, 2021).

Circular Economy in Turkey-Private

Instead of reporting, only the biggest firms' CE applications in Turkey, we decided to focus on the biggest groups' companies and we have identified these firms using the "Fortune 500" largest private company rankings (net sales) in Turkey. The main idea is to understand how these big groups perceive and apply CE concepts in their businesses. We believe that these leading groups' practices in the application of CE concepts set an example to all followers in many sectors. Six groups, Koç, Sabancı, Ahlatçı, Kazancı, Anadolu, Zorlu have been identified. Their operations include manufacturing, energy, mining, technology, retail, construction, automotive, communication, agriculture, petro chemicals and transportation. These six groups have thirty-seven companies in the Fortune 500 (Group, Medya, 2020) list, their total net sales add up to more than 25 percent of the total sales of the companies in the Fortune 500 list. Upon identifying these leading groups in Turkey, we investigated their applications of Circular Economy concepts in their businesses through their websites and reports. Table 3.1 lists the selected groups' companies listed in the Fortune-500. Below, a brief description of these companies' activities regarding their applications of the sustainability and circular economy concepts is provided.

Koç Group

There are seventeen Koç Group companies in the Fortune 500, ranking between 2nd and 429th. Koç companies' total sales is over 14 percent of the list's total.

Koç Group's Environmental Management principles are determined in line with UN Global Compact (UNGC) and UN Sustainable Development Goals (SDGs). These principles include prevention of waste generation, reducing waste, increasing recycling, recovery, reuse, using waste as a resource whenever possible in line with the principles of CE. In addition, the group aims to provide a role model for other businesses within the group and outside the group. The group brings environment, energy and sustainability experts from all Koç Group companies and form the Koç Environment Committee. The Committee coordinates all efforts (Koç, 2021). However, each company is independent of taking actions according to the needs of their respective sector (Koç, 2022). Next, some of the individual practices and accomplishments of Koç Group companies are summarized.

As Turkey's largest industrial establishment and Europe's seventh largest refining company, Türkiye Petrol Rafinerileri A.Ş. (Tüpraş), according to its "Strategic Transformation Plan" aims to reduce its carbon emissions by 27 percent by 2030 compared to 2017, and become carbon neutral by 2050. In 2019, Tüpraş saved 473,234 GJ of energy because of its energy-efficiency projects, which is equivalent to 26,292 tons of CO_2 (Koç, 2022). Tüpraş general manager İbrahim Yelmenoğlu states that Tüpraş will focus on Sustainable Refining, Zero-Carbon Electricity and Green Hydrogen in the near future. To accomplish this energy transformation goal, Tüpraş will set aside $50 million for venture capital investments until 2030 (Tüpraş, 2022). Between 2014 and 2015, in collaboration with the

Table 3.1. Selected Groups, Companies and Their Net Sales in Fortune 500, 2020

Rank	*Name of the Company*	*Name of the Group*	*Net Sales*
2	TÜRKİYE PETROL REFINARY	KOÇ GROUP	63.243.815.000 ₺
4	FORD AUTOMOTIVE	KOÇ GROUP	49.451.407.000 ₺
7	AHLATCI JEWELRY	AHLATÇI GROUP	41.484.599.264 ₺
8	ARÇELİK	KOÇ GROUP	40.872.483.000 ₺
9	OPET	KOÇ GROUP	38.623.155.000 ₺
11	AHLATCI METAL REFINERY	AHLATÇI GROUP	31.470.037.405 ₺
12	TURKCELL COMMUNICATIONS SERVICES	KOÇ GROUP	29.103.738.000 ₺
13	MİGROS	ANADOLU GROUP	28.790.190.000 ₺
16	ANADOLU EFES BEER AND MALT	ANADOLU GROUP	26.742.693.000 ₺
18	ENERJİSA	SABANCI GROUP	21.757.203.000 ₺
20	VESTEL	ZORLU GROUP	21.546.180.000 ₺
27	OTOKOÇ AUTOMOTIVE	KOÇ GROUP	14.259.793.000 ₺
37	AYGAZ	KOÇ GROUP	10.144.743.000 ₺
41	ENERJİSA ÜRETİM SANTRALLERİ A.Ş.	SABANCI GROUP	9.345.371.000 ₺
42	ZORLU ENERGY	ZORLU GROUP	8.579.113.000 ₺
46	CARREFOURSA	SABANCI GROUP	7.914.987.337 ₺
50	AKSA ENERGY.	KAZANCI GROUP	7.230.546.535 ₺
51	AKSA NATURAL GAS	KAZANCI GROUP	7.117.733.879 ₺
55	TÜRKTRAKTÖR	KOÇ GROUP	6.243.308.297 ₺
58	TEKNOSA	SABANCI GROUP	5.606.519.000 ₺
73	KORDSA	SABANCI GROUP	4.536.336.439 ₺
74	BİLKOM	KOÇ GROUP	4.433.401.790 ₺
78	ÇELİK MOTOR	ANADOLU GROUP	4.282.454.000 ₺
82	BRİSA	SABANCI GROUP	4.236.875.447 ₺
93	ZER MERKEZİ HİZMETLERİ VE TİC. A.Ş.	KOÇ GROUP	3.944.499.133 ₺
97	RAM DIŞ TİCARET A.Ş.	KOÇ GROUP	3.751.203.012 ₺
124	OTOKAR AUTOMOTIVE & DEFENSE	KOÇ GROUP	2.908.711.000 ₺
129	AKSA ELECTRIC	KAZANCI GROUP	2.831.538.177 ₺
172	ÇİMSA	SABANCI GROUP	2.076.298.962 ₺
177	KOÇ SYSTEMS	KOÇ GROUP	2.036.134.977 ₺
178	THY OPET AVIATION FUEL	KOÇ GROUP	2.033.944.746 ₺
187	AKÇANSA	SABANCI GROUP	1.967.108.077 ₺
228	AKSA POWER GENERATOR	KAZANCI GROUP	1.602.781.917 ₺
254	SETUR	KOÇ GROUP	1.449.068.391 ₺
262	DİTAŞ	KOÇ GROUP	1.384.308.000 ₺
284	ANADOLU ISUZU AUTOMOTIVE	ANADOLU GROUP	1.241.213.253 ₺
429	TAT	KOÇ GROUP	802.944.100 ₺
	Fortune 500 Total		**1.938.083.036.239 ₺**

Source: https://www.fortuneturkey.com/fortune500

Municipality of Körfez, Tüpraş initiated the Urban Wastewater Recycling Project and invested $24.8 million in İzmit Refinery. In four years after the project started, 87 million cubic meters of wastewater was recycled. In 2019, Tüpraş İzmit Refinery became the first refinery to use the Absorption Heat Exchanger (AHT) system, "Indus3Es" project, (was developed under Horizon 2020) which recycles low temperature (below 130 degrees Celsius) waste heat (Koç Group, 2022).

Ford Otosan as the leading company in automotive sector in Turkey, which also operates the largest commercial vehicle production compound in the European region, aims to reach net zero emissions by 2050. Before reaching its ultimate goal, the company is expecting to reduce carbon emissions per vehicle by 55 percent by 2030 and switch to zero-emissions in trucks by 2040. In order to achieve its goal, the company has focused on the production of electric vehicles and will have invested 20.5 billion TRY by mid-2026 to expand its plants in Kocaeli to produce a new technology battery with 50 percent increased energy intensity, 25 percent reduced charging time, 20 percent reduced weight, and extended lifetime. Moreover, Ford Otosan and Otokar's long-range hydrogen fueled bus with fuel cell technology project application for Horizon 2020 funding was accepted.

Arçelik, as one of the biggest durable consumption goods manufactures in Turkey ranked at the eighth position in the Fortune-500 list. The company's approach to sustainability issues and business efforts is interlinked. As a result, Arçelik manages most of its wastes and uses them as an input in the new production processes. In 2019, the Ministry of Environment and Urbanization awarded Arçelik with "Zero Waste Private Sector Award" as a result of the company's success in waste management and recycling facilities (Koç, 2022). In 2021, the company improved its score in both Climate Change and Water Security Programs of Carbon Disclosure Project (CDP) to an "A–."

Aygaz A.Ş., as the first company to publish its sustainability report in 2002, obtained the Zero Waste Certificate for the Dörtyol facility in 2020. The company also aims to recover as much water as possible and over the three years' period the company managed to reduce its water consumption by 14 percent. In 2020, upon implementing energy efficiency projects across Aygaz, the company's energy savings reached to 25,000 kW in electricity and 321,000 kWh in natural gas and CO_2 emission was reduced by 108 tons (Aygaz, 2022).

Other Koç Group companies either do not have any information on their web sites in regards to their applications of CE concepts or are just following the guidelines established by the Koç Group.

Sabancı Group

In the Fortune 500 list, Sabancı Group has eight companies operating in various sectors, such as energy, cement, retail, and industrial. Their total sales contribute to almost 3 percent of the total of the Fortune 500 list. The Group claims to comply all legal requirements and environmental regulations. In addition, they claim to have set and implement targets for energy and waste management and natural resource consumption (Sabancı, 2022).

As the largest electricity distribution and retail sales company in Turkey, Enerjisa claims to be a pace setter in the energy sector. The company has plans and actions in various fields including carbon emission reduction, increasing energy efficiency, conservation of biodiversity, ecosystem risk assessment, tackling climate change, and transiting to CE in waste management. For clean and more efficient energy distribution, the company invested in materials in network structure in the amount of 1.8 billion TRY in 2020 (Enerjisa, 2022a). In 2010, Enerjisa initiated "I protect my energy" project. The project aimed to inform and educate kids about the importance of energy efficiency and conservation using a mobile game. The goal of the project is to raise awareness of 1 million children on energy efficiency. The project was granted four awards, in 2017 the "Social Responsibility Program of the Year in Europe," in 2018 the "Responsible Consumption and Production Award," "The Best Energy Management Award," as well as the "Golden Compass Award" (Enerjisa, 2022b). Enerjisa bought the majority of the shares of Eşarj, which is the largest electric vehicle-charging operator in Turkey. In 2018, Eşarj became the first operator in Turkey to obtain the International Renewable Energy Certificate (IREC) by using renewable energy in all its stations. The company is expected to reduce its carbon footprint by nearly 30,000 tons per year (Enerjisa, 2022c). With its "Energy Efficiency Project," Enerjisa aims to decrease general energy consumption by doing necessary adjustments, recycling waste energy and optimizing the existing equipment in qualified businesses. Enerjisa offers its services under the "Energy Performance Contract" (EPC) model. Upon doing necessary inspections at the candidate firm, Enerjisa will guarantee the performance of the project and up to 80 percent reduction in energy costs can be achieved (Enerjisa, 2022d).

Carrefoursa has been operating as one of the leading brands of the organized food retail industry in Turkey. In order to implement "the zero waste system" in 2021, the company established waste collection centers in approximately 100 stores around Turkey. Any waste with a value—such as paper, metal, plastic, vegetable oil, battery, etc.—was collected separately and sent to licensed recycling companies to contribute to their disposal. As a result, in 2019, 1,486 tons of plastic, 359 tons of paper, sixty-one tons of glass and sixty-five tons of metal materials in self-branded products were recycled and brought into the environment. In addition, approximately fifteen tons of waste oil was collected from stores and converted into biodiesel (Carrefoursa, 2022a). As of 2022, Carrefoursa was able to get 230 of its stores "zero waste certified" around Turkey (Carrefoursa, 2022b).

Kordsa is one of the leading companies in tire and construction reinforcement and composites markets in the world and operates in twelve facilities throughout five countries (Kordsa, 2022). In 2020, the company managed to save 2.7 million USD from the reuse of 2.4 million pieces of materials and invested over one million USD environmental improvements. Kordsa has also decreased its emissions by 9.1 percent and recycled 22.4 percent of its wastes. Being an internationally recognized firm, Kordsa acted faster than most of the other companies in Sabancı Group and signed the UNGC in 2014 and prepared Carbon Disclosure Project (CDP) Supply Chain report. In 2016, Kordsa started preparing CDP Climate Change, Water Security and Supply Chain Reports. In 2017, CDP Climate Change score was D and Water Security score was C. In three years, the company raised its CDP Climate Change and Water Security scores to A-. Since 2014, Kordsa has turned the by-products of Nylon 6.6 yarn production facilities into pellet forms. This new output can now be used as zero emission raw material input for engineering plastics industry. As a result, the company's sales have increased from the sale of this by-product, manufacturers benefited from the product being supplied locally and its imports declined and resulted in higher profits for the buyers and in a global scale the company has gained about three thousand tons of polymer waste to the circular economy (Kordsa, 2020).

Brisa, the leader of the Turkish tire industry, according to its climate change policies, aims to reduce its direct emissions by 56 percent in 2030, achieve net zero emissions in 2050, and complete its low carbon emission transition (Brisa, 2021).

Çimsa and Akçansa are the largest cement producers in Turkey. Çimsa has been preparing its Climate Change Report since 2014 and Water Security Report since 2015. The company's score in Climate Change has improved from a C to a A- where its Water Security score has improved from a B- to a B. In 2020, Çimsa established the Zero Waste Management System at its five plants and has been awarded the Zero Waste Certificate. The company's environmental investments increased by 18.63 percent in 2020 and reached to 6.76 million TRY. In its combat with climate change and green gas emissions, Çimsa has been trying to adopt the use of alternative fuels. The Company is expected to increase its alternative fuel usage to 23 percent and reduce the CO_2 ratio for every ton unit of clinker to 1.95 percent by 2025 (Çimsa, 2020).

Akçansa places energy efficiency, climate change, greenhouse gas emissions, low environmental impact products, water, and waste management among their very high priorities. Similar to Çimsa, Akçansa is also converting to alternative and renewable energy sources to manage their greenhouse gas emissions. For instance, between 2014 and 2019, Akçansa used one million tons of Istanbul's waste as an alternative source of fuel and raw materials. As a result, the company reduced greenhouse gas emissions per unit turnover by 24 percent between 2017 and 2020. Twelve percent of the company's electricity energy comes from renewable sources such as wind and waste heat. In 2019, 85 percent of the water used in the company was recycled and reused as cooling water (Akçansa, 2020).

Ahlatçı Group

Ahlatçı Group companies operate in six different industries with forty-three companies; The group's major fields of activity are refinery and jewelry sectors. Even though there are only two Ahlatçı Group companies, Ahlatçı Jewelry and Ahlatçı Metal Refinery, in the Fortune 500 list, their total sales contribute to more than 3.7 percent of the Fortune 500's total. Unfortunately, we were not able to find any information in regard to these companies' sustainability or circular economy concept applications in their web site, except they claim to comply with national and international legislations and regulations in terms of efficient use of energy and natural resources to prevent any negative environmental effects (Ahlatçı, 2022).

Zorlu Group

Although Zorlu Group's companies only contributed 1.55 percent of the Fortune 500 lists' total sales, the group is among the top three groups in Turkey in terms of employment creation and one of Turkey's seven big exporters (Zorlu, 2021). With the implementation of new technologies to improve energy efficiency and procedure optimizations, in 2020 Zorlu

Group reduced its carbon emissions by 91,339 CO_2 tons. Compared to 2019's energy savings, in 2020 energy savings tripled, thanks to newly adopted technologies and procedures. "Waste paint recycling and usage system," where paint particles used in the production process can be recycled and used again in the production, was developed by Vestel. This system was first used by Vestel Electronics and has received three patents. Using the system, a total of 177 tons of waste paint is recycled and CO_2 emissions reduced by 460 tons and 123 tons of water consumption was avoided. Zorlu Energy on the other hand supports the transition to electric vehicles and since 2018, the company has installed charging stations in every city in Turkey. Moreover, these charging stations are certified with Renewable Energy Certificate (I-REC).

Kazancı Group

Four of the Kazancı Group Companies have been ranked among the Fortune 500 companies. Among them, Aksa Energy claims to monitor sustainability platforms on local and global scale. The company's voluntary support in sustainable development in all areas has let it be included in the Borsa Istanbul (BIST) Sustainability index since 2015. The company has also been a signatory of the United Nations Global Compact (UNGC) since 2017 (Aksa, 2022). UNGC supports companies to do business responsibly by aligning their strategies and operations with Ten Principles,[3] and take strategic actions to advance societal goals such as UN Sustainable Development Goals (UNGC, 2022). Aksa Energy established its Sustainability Committee in 2015 which contributes to seven of the UN Sustainable Development Goals. As many of the companies in the list, Aksa Energy bases its environmental policy on four main pillars: climate change, natural resource management, waste management and conservation of biodiversity. Between 2018 and 2020, the company increased the amount of recycled wastewater by 3.27 times and its consumption of municipal and ground water has significantly decreased, 75 percent and 88 percent respectively. The company claims to be committed in the battle against climate change and therefore signed the "Trillion Tonne Communique"[4] (Aksa Energy, 2020).

In 2020, Aksa Natural Gas, allowed households to replace coal with natural gas. As a result, carbon emissions have decreased by seven million tons in its license regions. In 2019 Kazancı Group companies joined forces for a social responsibility project, "Energy for tomorrow" which entails a team of corporate communication employees to travel all around Turkey to educate middle school students about every aspect of energy, such as energy resources, energy efficiency, energy literacy, clean energy, etc (Kazancı Holding, 2020).

Anadolu Group

The group is currently operating in beer, beverage, retail, agriculture, stationary, automotive, quick service restaurant, and energy sectors with nearly eighty companies and more than 80,000 employees in nineteen countries including Turkey. Four of these companies are ranked among the Fortune 500 companies and these companies contributed to 3.15 percent of the total sales of the list.

According to the group's 2015–2020 Sustainable Development Compatibility Report, out of 555 projects brought to life 29 percent of them were related to environmental issues, and 66 percent of them were related to social matters (Anadolu Grubu, 2020). Migros, one of the biggest retail companies in Turkey was first acquired by Koç Group in 1975 and in 2015, the majority of its shares were transferred to Anadolu Group (Migros, 2021) which ranked 13th in Fortune 500 list. The company scored an "A-" with its proposed tackling climate change reports to CDP in 2016, 2017, and 2019, and placed itself among Turkey's Climate Leaders. One year later, Migros was among the CDP Water Program Leaders, too. Between 2015 and 2020, Migros reduced its carbon emissions by 5.1 percent per square meter daily. Its patented 100 percent natural chiller cooling system emits almost zero CO_2. The company reduced water consumption by 22.9 percent by educating its staff. Migros aims to reduce its electricity consumption by 18.5 percent by 2024 compared to 2013. However, the company exceeded this goal and reduced its electricity consumption by 24.2 percent. On top of that, using cloud technology the company managed to save 80 percent in energy. In 2020, Migros collected thirty-one tons of waste vegetable oil, sent 5.68 tons of batteries for recycling, and collected 77.38 tons of electrical and electronic waste (Anadolu Grubu, 2020).

Ranked 16th, Anadolu Efes is Europe's fifth and the world's ninth largest brewer by production volume (Anadolu Efes, 2022). Between 2015 and 2020 Anadolu Efes managed to reduce its cost of production through improvements in efficiency and saved 611,000 USD. These savings were in the equivalent of 6,000 individuals' annual electricity consumption of 21,739 MWh of energy. In the same period, the company saved 288,000 cubic meters of water. The company also changed the material of its disposable glasses from the ones that dissolve in 450 years in the nature to the biodegradable ones that dissolve and leave no residue only in two years. Through waste management efforts, the company prevented 18,000 tons of waste and managed to get "Zero Waste" certificate for all malt and beer production facilities in Turkey (Anadolu Grubu, 2020). Unfortunately, we could not find any significant information regarding Çelik Motor's and Anadolu Motor's circular economy implementations.

Conclusion

Early days of civilization where everything was scarce and mass production and extraction of natural resources were either very limited or even impossible, people would follow today's circular economy concepts in every aspect of their lives because these concepts were actually the laws of the nature. With industrial evolution and advancements in technology, urbanization, and especially elevated borders and easier transportation opportunities at first made everything abundant but when the limits of our big blue sphere were reached we realized the laws of the nature and now try to implement these concepts to our lives to remain on this earth a little longer.

In conclusion, every major group in Turkey seems to place energy efficiency, climate change, greenhouse gas emissions, low environmental impact products, water and waste management among their very high priorities. Only a few have established industrial symbiosis and complete circularity in their organizations, which may very well be due to some technical and cost related issues. However, unless companies have international ties or marketing concerns, they do not want to take any chances and change their way of manufacturing. Such companies believe that going through such transitions may result in loss of customers. Using recycled or restored materials may seem like a reduction in the general quality of the merchandise. The government's role in this situation is very critical and both customers and manufactures should be enlightened about this misinformation. Some of the companies mentioned above understood and experienced the benefits of this transition and gained a leading position in their sectors (Güngör, 2019). As Env.net (2020) points out, waste management seems like the most important CE item in public policy agenda in Turkey. Major groups' companies have relatively easy access to information and by using their resources they manage to adopt CE principles. Small and mid-sized companies, public authorities, nongovernmental organizations are still relatively uninformed and reluctant to take action in circularity movement.

Awareness should be established by the coordination and strategic planning of government authorities, supported by the leaders of the business sectors, optimized and improved by the knowledge and research and development activities of the academia and finally passed on to the new generations through their family values.

Notes

1. Which can be interpreted as Industrial Symbiosis, the use of a factory's wastes and excess energy by another factory.
2. (KPMG, 2022)
3. See (UNGC, 2022)
4. "*The Trillion Tonne Communique is a global call to arms from businesses who take the science of climate change seriously and are demanding a proactive policy response.*" (University of Cambridge, 2014)

References

Ahlatçı. (2022). *Sürdürülebilirlik ve Çevre*. Retrieved May 1, 2022, from Ahlatçı Web Site: https://www.ahlatci.com.tr/surdurulebilirlik.html.

Akçansa. (2020). *Sustainability Report 2018–2019*. Retrieved April 30, 2022, from Akçansa Web Site: http://www.akcansa.com.tr/docs/Sustainability-Report-2018-2019.pdf.

Aksa. (2022). *Our Approach to Sustainability*. Retrieved April 30, 2022, from Aksa Energy: https://www.aksaenerji.com.tr/en/sustainability/our-approach-to-sustainability/.

Aksa Energy. (2020). *Sustainability*. Retrieved May 2, 2022, from Aksa Energy Web Site: https://www.aksaenerji.com.tr/en/sustainability/.

Anadolu Efes. (2022). *About Us*. Retrieved April 10, 2022, from Anadolu Efes Web Site: https://www.anadoluefes.com/en/sayfa/1/438/chairmans-message.

Anadolu Grubu. (2020). *Sürdürülebilirlik*. Retrieved April 22, 2022, from Anadolu Grubu: https://www.anadolugrubu.com.tr/Upload/Docs/SKA_Uyumluluk_Raporu_2015_2020.pdf.

Aygaz. (2022). *Sustainability Reports*. Retrieved April 30, 2022, from Aygaz Web Site: https://www.aygaz.com.tr/uploads/kurumsal/surdurulebilirlik/13a154a4_4639_448f_9439_b33e7064a86b__skr_en_web.pdf.

Brisa. (2021, September 21). *Haberler*. Retrieved May 3, 2022, from Brisa Web Site: https://www.brisa.com.tr/haberler/Brisa-n%C4%B1n-emisyon-azalt%C4%B1m-hedefleri-bilim-temelli-hedefler-giri%C5%9Fimi%E2%80%99nden-onay-ald%C4%B1.

Carrefoursa. (2022a). *Çevresel Çalışmalarımız*. Retrieved May 1, 2022, from Carrefoursa Web Site: https://www.carrefoursa.com/kurumsal/surdurulebilirlik/cevresel-calismalarimiz/.

Carrefoursa. (2022b). *Sürdürülebilirlik Bülteni-4*. Retrieved May 1, 2022, from Carrefoursa Web site: https://www.carrefoursa.com/kurumsal/wp-content/uploads/2022/04/CSA-Su%CC%88rdu%CC%88ru%CC%88lebilirlik-Bu%CC%88lteni-4.-Sayi.pdf.

Çimsa. (2020). *Sustainability Reports*. Retrieved May 1, 2022, from Çimsa Web Site: https://www.cimsa.com.tr/ca/docs/233E1C84A98E461EB5C6547C6B0972/BAA5D5811DAB49B4AB02EA3ACB4A4F70.pdf.

Ellen Macarthur Foundation. (2017). *Archive*. Retrieved May 6, 2022, from Ellen Macarthur Foundation Web Site: https://archive.ellenmacarthurfoundation.org/explore/the-circular-economy-in-detail.

Enerjisa. (2022a). *Environmental Management*. Retrieved May 1, 2022, from Enerjisa Web Site: https://www.enerjisa.com.tr/en/sustainability/role-model-within-the-sector/environmental-management.

Enerjisa. (2022b). *Projects*. Retrieved May 1, 2022, from Enerjisa Web Site: https://www.enerjisa.com.tr/en/sustainability/projects/with-the-i-protect-my-energy-project-energy-conservation-is-taught-by-gaming.

Enerjisa. (2022c). *Eşarj Breaking Ground Again with International Renewable Energy Certificate (IREC)*. Retrieved May 1, 2022, from Enerjisa Web Site: https://www.enerjisa.com.tr/en/sustainability/news-and-awards/esarj-breaking-ground-again-with-international-renewable-energy-certificate-irec.

Enerjisa. (2022d). *Energy Efficiency*. Retrieved May 1, 2022, from Enerjisa Web Site: https://www.enerjisa.com.tr/en/energy-of-my-business/energy-efficiency.

Env.net. (2020). *Circular Economy Country-Specific Report Written Contribution for the 2020 Annual Report Turkey*. İstanbul: TEMA.

European Commission. (2019). *A European Green Deal*. Retrieved April 15, 2022, from European Commission Web Site: https://ec.europa.eu/info/strategy/priorities-2019-2024/european-green-deal_en#highlights.

Group, Medya. (2020). *Fortune 500*. Retrieved April 20, 2022, from Fortune Türkiye: https://www.fortuneturkey.com/fortune500.

Güngör, E. (2019). *To Cycle or not to Cycle Towards a Circular Economy in Turkey*. Netherlands Enterprise Agency.

Hedefler İçin İş Dünyası. (2020). *Döngüsel Ekonomi Rehberi*. Retrieved April 20, 2022, from business4goals: https://business4goals.org/PDF/Dongusel_Ekonomi_Rehberi.pdf.

K.Winans, A.Kendall, and H.Deng. (2017, February). The history and current applications of the circular economy concept. *Renewable and Sustainable Energy Reviews*, 68(1), 825-833.

Kazancı Holding. (2020). *Annual Report 2020*. Retrieved May 1, 2022, from Kazancı Holding Web Site: https://www.kazanciholding.com.tr/assets/Kazanci_Holding_Annual_Report_2020.pdf.

Kirchherr, J., Reike, D., and Hekkert, M. (2017, December). Conceptualizing the Circular Economy: An Analysis of 114 Definitions. *Resources, Conservation and Recycling*, 127, 221-232. doi: https://doi.org/10.1016/j.resconrec.2017.09.005.

Koç. (2021). *Koç Group Environmental Policy*. Retrieved April 18, 2022, from Koç: https://cdn.koc.com.tr/cmscontainer/kocholding/media/koc/05surdurulebilirlik/policies/environmental-policy.pdf.

Koç. (2022). *Addressing Climate Change Impacts*. Retrieved April 15, 2022, from Koç: https://www.koc.com.tr/sustainability/act-for-the-planet-together/change-climate.

Koç. (2022). *Circular Economy and Tackling Plastic Wastes*. Retrieved April 15, 2022, from Climate Change: https://www.koc.com.tr/sustainability/act-for-the-planet-together/change-climate#circular-economy-and-tackling.

Koç. (2022). *Climate Change*. Retrieved April 15, 2022, from Koç: https://www.koc.com.tr/sustainability/act-for-the-planet-together.

Koç Group. (2022). *Reports and Policies*. Retrieved April 30, 2022, from Koç Group Web Site: https://cdn.koc.com.tr/cmscontainer/kocholding/media/koc/05surdurulebilirlik/policies/koc-group-sustainability-report-2020.pdf.

Kordsa. (2020). *2020 Sustainability Report*. Retrieved May 1, 2022, from Korda Web Site: https://www.kordsa.com/en/images/pdf/2020_Sustainability_Report.pdf.

Kordsa. (2022). *Kordsa Web Site*. Retrieved May 1, 2022, from About Us: https://www.kordsa.com/en/kordsa/detail/about-us/3/1/0.

KPMG. (2022). *Preliminary Research on Circularity Potential of Five Sectors in Turkey*. AFD. Retrieved from http://www.skdturkiye.org/images/icerik/Preliminary_Research_Report_on_Turkeys_Circular_Economy_Potential_in_5.pdf.

Migros. (2021). *Our History*. Retrieved April 10, 2022, from Migros Kurumsal: https://www.migroskurumsal.com/en/about-us/our-history.

Nuroğlu, E. (2019). *Döngüsel İktisat Yolunda Türkiye: Sıfır Atık Projesi*. Retrieved May 7, 2022, from AA: https://www.aa.com.tr/tr/analiz/dongusel-iktisat-yolunda-turkiye-sifir-atik-projesi/1520219.

Potting, J., Hekkert, M., and Worrell, E. (2017). *Circular economy: Measuring innovation in the product chain-Policy Report*. The Hague: PBL Publishers.

Sabancı. (2022). *Environment*. Retrieved May 2, 2022, from Sabancı Web Site: https://www.sabanci.com/en/social-responsibility/our-social-responsibility-activities/environment.

Salihoğlu, G., Turhan, Ş., İpeker, G., and Salihoğlu, N. (2021). Circular Economy in Turkey. In S. G. Ghosh (Ed.), *Circular Economy: Recent Trends in Global Perspective* (pp. 527-572). Singapore: Springer. doi: https://doi.org/10.1007/978-981-16-0913-8_17.

T.C. Ticaret Bakanlığı. (2021). *Yeşil Mutabakat Eylem Planı 2021*. Ankara: T.C. Ticaret Bakanlığı. Retrieved from https://ticaret.gov.tr/data/60f1200013b876eb28421b23/MUTABAKAT%20YE%C5%9E%C4%B0L.pdf.

Tüpraş. (2022). *Press Releases*. (Tüpraş, Producer, and Press Releases) Retrieved April 20, 2022, from Tüpraş: https://www.tupras.com.tr/en/press-releases?detay=press-relaese-112021.

UNGC. (2022). *The Ten Principles of the UN Global Compact*. Retrieved April 30, 2022, from United Nations Global Compact: https://www.unglobalcompact.org/what-is-gc/mission/principles.

University of Cambridge. (2014). *Resources*. Retrieved May 1, 2022, from University of Cambridge Web Site: https://www.cisl.cam.ac.uk/resources/low-carbon-transformation-publications/trillion-tonne-communique.

Zorlu. (2021). *Corporate*. Retrieved April 26 2022, from Zorlu Web Site: https://www.zorlu.com.tr/en/corporate/zorlu-with-numbers.

Circular Economy Approach in Automotıve Aftermarket Products and Services

Revision Spare Parts Solutions in Turkey

AYŞE GÖNÜL DEMIREL AND VEYSEL KARANI NURDAĞ

4

Introduction

In the changing after-sales spare market, finding innovative solutions that will minimize the environmental impact of waste spare parts and bring them back to the economy is considered as an important issue in the literature. It is an important issue for companies in the after-sales spares market to re-introduce waste spare parts to the economy and to develop solutions that can help reduce negative environmental impacts (http://www.osd.org.tr/). There are several strategic ways to acquire spare parts with reuse opportunity at different levels (such as the end-of-life strategy). Such strategies allow optimizing the product life cycle and reducing environmental impacts. These strategies offer great opportunities for the recovery of used products and parts after the sale (Saavedra et al., 2013).

According to Giuntini and Gaudette (2003) and the resource-based view, a company can offer better opportunities to increase the productivity and profitability of remanufacturing by pointing to the preservation of some raw materials and added value during the manufacture of the product (Giutini and Gaudette, 2003). Remanufacturing is an industrial process in which used products, called cores, are returned to their useful life. Remanufacturing, which can be considered as a kind of reverse logistics activity, is a process that includes the systematic planning, in-process inventory, implementation, and control of the flow of raw materials, packaging and finished products from the point of production, distribution or use to the appropriate disposal point or recovery point (Subramoniam et al., 2009). Remanufacturing, recycling, repair, reuse, refurbishment, and refurbishment differ from other end-of-life recoveries, but they can use the same processes or end up with the same end product. Remanufactured products have better-quality demonstrations by going through the recycling and reconditioning process (Armacost et al., 2002)

Remanufacturing is a manufacturing strategy that allows more material to be recycled than normal recycling methods (Güngör and Gupta, 1999). With the remanufacturing processes, the quality performance is kept at the optimum level, the functionality of the product is preserved and productivity increases by contributing to the circular economy (Nasr and Thurston, 2006). In addition, automotive aftermarket spare parts activities include providing "original spare parts" and "OES equivalent spare parts" after the end of the warranty period in the "Independent Aftermarket Market" (IAM) (Subramoniam et al., 2009). Remanufacturing plays an important role for OEMs. Remanufactured products are considered as an important alternative in customer perception with their performance features, warranty, and other features. (Ijomah et al., 2007). This situation is seen as an important resource input for the producers to continue their existence in a sustainable and competitive way

According to a survey in the United States and Europe, there has been an increased awareness of business sustainability issues by all stakeholders in the value chain. Accordingly, it has emerged that businesses should attach importance to adopting sensitive policies that give importance to the environment and circular economy in all production processes. According to previous studies in the literature, it has been concluded that remanufacturing is a profitable and environmentally friendly production process. Remanufacturing provides not only economic benefits, but also ecological benefits, as it is a situation where a product seen as waste is reused (Subramoniam et al., 2009).

With the increasing need for spare parts in Turkey, the reman process attracts the attention of companies. Businesses contribute to the growth of the sector by investing more in this area and at the same time consider producing more environmentally friendly spare parts solutions as an important resource (https://www.taysad.org.tr/). It is accepted in the literature that integrating the remanufacturing strategies of enterprises in the after-sales market into all customer-oriented stages creates an important competitive advantage. Remanufacturing offers opportunities to increase aftermarket revenue potential and develop lower-cost opportunities. Optimized supply chain management is accepted in the literature as an important resource input for companies whose basic strategies are sensitive to both the environment and the circular economy. In the after-sales market, businesses attach importance to both reproduction and supply chain issues while determining the decision points of their strategies (Subramoniam et al., 2009).

In the automotive aftermarket, sustainability, providing better performance and designing environmentally friendly remanufacturing processes are explored in the literature as an important concept. Studies on the subject include a new field and concepts and are developing day by day (Yusopet al., 2016). An extensive literature review on Remanufacturing Design was conducted by Hatcher and others (2011). According to this literature review, it is seen that the studies focus on improving the design methods and improving the product design and focusing on the design of the parts suitable for reproduction.

When these studies are examined, it is seen that there is a lack of an in-depth study on the effect of organizational factors in order to make the design process suitable for reproduction. There is increasing interest in remanufacturing in different sectors. Examples of remanufactured and long-lasting products in the Automotive and Electrical-Electronics (EEE) sectors are copiers, washing machines, disposable cameras, ink cartridges, and so on similar products can be shown as an example (Hatcher et al., 2011). Remanufacturing case studies in sectors where different products are produced, such as machinery and mechanical tools, rubber, textiles, have been reviewed in the literature. According to these studies, the automotive aftermarket of remanufacturing includes the most case (Yusop et al., 2016). Automotive aftermarket is the sector with the highest participation in remanufacturing. The methods handled in this sector constitute 66 percent of the transaction volume of all remanufacturing methods (Steinhilp, 2001).

As explained in Seitz's study (2007), although it has ethical and moral responsibilities, remanufacturing is a production process that responds to the need for part replacement and economic cost gains, which occur due to the recall of the damaged product and legal warranty reasons, among other elements. Automotive spare part manufacturers use remanufacturing for different purposes. Some of these purposes are to eliminate customer dissatisfaction that occurs when spare parts supply is not guaranteed and to compete in the product group with low market share. When designing the remanufacturing process, businesses determine their strategies by considering cost savings, increasing the profit percentage, and environmental and legal obligations (Yusopet al., 2016).

Within the context discussed above the main purpose of this study is to introduce a new business model of remanufacturing in the automotive spare parts sector. In this study, remanufacturing will be investigated both theoretically and practically, and the result will be modeled through a case application by considering its basic features. The result of the study is an improved model of the remanufacturing process in the automotive aftermarket based on a set of market factors derived from the case study. In the study, each stakeholder using remanufactured parts in automotive aftermarket processes, including Independent Aftermarket (IAM) and authorized services, will be examined.

A review of the development of the reproduction literature, its benefits and the actors in the process has been made. Reproduction features, processes and actors were analyzed and defined for the case under consideration. While the concepts of reproduction and reverse logistics have gained popularity in practice, the maturity of the existing literature and theory has not reached a sufficient level in terms of creating a new business model in these areas in Turkey. In this study, a new business model has been developed specifically to fill the gap in automotive aftermarket of automotive remanufacturing. All the resources required for the study were reviewed and the value chain was reviewed. These are customer demands, product design and development, cost-benefit analysis, basic supply management, personnel competencies and skills, product lifecycle strategies, key stakeholders, environmental issues, regulations and impact of emerging economies, and reverse logistics network design.

With the findings obtained and the analysis of the customer experience in the market, the new business model was designed and verified. With the model, it is aimed to increase the business volumes in a way to meet the demands of the customers by using the remanufactured parts in the independent after-sales market of the authorized services more effectively.

Authorized service dealer shows that it has more advantages over IAM in terms of remanufacturing operation and marketing of remanufactured product. Collaboration between the remanufacturer and the dealership can be established to develop managerial practices, leading to better remanufacturing operations and higher profitability for both.

Literature Review

Remanufacturing, as a process, is a process that requires a high level of importance throughout the organization. In this process, damaged, repaired or replaced, cleaned parts are partially or completely processed. Although the emergence of remanufacturing is purely economic according to some literature studies, it is a development that can compensate for the ecological imbalance caused by rapid population growth and excessive consumption of resources and contribute to the cyclical economy.

Remanufacturing protects the environment from the harmful effects of waste materials by making the products that have expired become reusable. In addition, companies using the remanufacturing process will gain high profits in a short time with less energy and materials and low cost according to the RBV approach, while they will have the lowest price in the market and be competitive for new market opportunities thanks to the products they offer. Customers will have the opportunity to purchase products of the same new product quality at a lower price. In this respect, remanufacturing contributes to the circular economy by benefiting both producers and customers (Güner and Güner, 2020).

As a fundamental process, remanufacturing involves many uncertainties. The process begins with the dismantling of the used product. Used products are disassembled, parts are cleaned and quality checked, reusable parts and scrap parts are sorted. After the reusable parts are determined, some of them go through additional processes and some of them go directly to the assembly process. In the final step, the component parts are assembled, checked for quality and packaged for shipping (Ferrer and Whybark, 2001).

Since original spare parts manufacturers have the most information about those products, they also have the most accurate information about which methods should be applied to integrate the product into the remanufacturing process. Ford Motor Company supported the involvement of several private institutions in the remanufacturing process for spare parts. Ford commissioned a network of manufacturers authorized to remanufacture spare parts components. Realizing that it is an effective way to procure spare parts remanufactured in this way, other automobile manufacturers followed the same path (Ferrer and Whybark, 2001).

A decision made at any stage of reproduction greatly influences the decision made at a later stage. This in turn affects KPIs such as efficiency, economic performance, usability and the amount of product recovered. Therefore, the need to improve the reproduction process at different stages of the decision is taken into account according to the KPIs (Ljunglöf and Nisser, 2017).

End of Life Process and Renewal of Product

According to the studies in the literature, when the renewal of products and their components is evaluated in terms of remanufacturing and product life cycle, it is concluded that remanufacturing strategies should be supported by businesses in order to obtain environmental and economic values. Reuse, refurbishment/refurbishment, repair, shredding, recycling and remanufacturing are among the most used and are briefly described below (Saavedra et al., 2013):

a) According to Gray and Charter (2007) and Rose (2000), Reuse is:
    i. The products are used immediately after the first using cycle and do not require any kind of restoration and repair. Potential problems can be protected obtained during initial use. There are no guarantees for products.
    ii. Type of raw material is used products and components
    iii. Level of Disassmbly: No
b) According to Gray and Charter (2007) and Ijomah et al. (2007), Repair is:
    i. Only damaged components of products are replaced in order to maintain product functionality. The warranty is only for the replaced components.
    ii. Type of raw material is used products and components
    iii. Level of Disassembly: Partial
c) According to King et al. (2006) and Rose (2000), Recycling is:
    i. Some used products reuse materials. In this case, static strength, structure and functionality are lost as well as the geometry of the original product.
    ii. Type of raw material is used products and components
    iii. Level of Disassembly: Total
d) According to Thierry et al. (1995), Cannibalizing is:
    i. Recovery of used parts and product quality depend on sub-parts to be obtained from a previously used or damaged part.
    ii. Type of raw material is used products
    iii. Level of Disassembly: Selective and Total
e) According to Gray and Charter (2007) and Ijomah et al. (2007), Refurbishment/Reconditioning is:
    i. The product and its components are made suitable for normal conditions of use. The warranty for a refurbished product is shorter than for a newly manufactured product. Used for technological upgrades.

ii. Type of raw material is used products and components
iii. Level of Disassembly: Partial

f) According to Gray and Charter (2007) Remanufacturing is:

i. Recovery of used parts and/or product. With a new product warranty and have the same quality features.
ii. Type of raw material is used products and components
iii. Level of Disassembly: Total

Remanufacturing

The aim of remanufacturing is to bring refurbished products to a level that meets quality standards and meets customer needs. Used products are completely disassembled and all sub-parts and components are thoroughly checked. Old or damaged components and modules are replaced with new ones. Repairable parts and sub-components are then assembled and thoroughly tested. Remanufacturing can be combined with an upgrading of the product. For example, a widely used machine tool can be upgraded with high-tech components that change after going through a remanufacturing process.

For example, BMW engines have been remanufacturing some high-value components of the engine for several years. Remanufactured engine components are tested according to the quality standards set by the brand. Remanufactured parts are sold with warranty and are 30–50 percent cheaper than new parts (Thierry et al., 1995). There are innumerable terms similar to "remanufacturing": Concepts such as reuse, refurbishment, repair, and so on, which are easily and commonly confused with "remanufacturing" due to the similarities in the process and the end product, are used in the automotive spare parts industry. In the past ten years, literature and research have discussed and studied these concepts.

Uncertainty, the limits of remanufacturing and as the volume of remanufacturing continues to increase, the factors creating uncertainty increase. If "recycling" were understood in the sense of "remanufacturing," it would be difficult to logically relate remanufacturing to a successful business model. The concept of remanufacturing has not yet effectively spread to manufacturing and related industries (Gray and Charter, 2007). When determining whether a product is remanufactured, it is important to consider the design of the process used. Remanufacturing is the process of reclaiming the material that added value when the product was first produced. Used components and spare parts are considered reproducible from the customer's point of view and are seen as equivalent to the original product (Ijomah, 2007).

To understand the factors and design of remanufacturing, it is worth noting that Remanufacturing can be divided into two categories:

1. Spare parts eg. starter motor, Automotive clutch,
2. Other components eg. Auto CR Scanner

This distinction is necessary to understand the differences that remanufactured parts and components can make in terms of product and design. All of the remanufacturing activity that makes up the parts industry is focused on market demand first; this is especially common in the automotive industry. On the other hand, it constitutes an important resource input for remanufacturing in legislation and processes (Gray and Charter, 2007).

Remanufacturing process consists of eight steps. Here is a basic sort, but may vary depending on the sequence of steps remanufactured products (Sundin, 2004):

1. Core collection
2. Investigation and identification of faults
3. Dismantling of all products
4. Cleaning of all parts (and storage)
5. Parts of renewal (and replaced with new parts if necessary)
6. Reassembly products
7. Testing to verify that serves as a new product

"Core" is a term used to describe the product or component to be widely performed during the reproduction process.

One of the most important resource inputs of remanufacturing is the design of parts and components. Parts with suitable design are more preferred for remanufacturing. Parts whose design is not suitable for reprodution are not feasible and cannot be considered as recycling. (Sundin, 2004).

According to the US Automotive Parts Manufacturers Association (ARPA), remanufacturing is defined as the process of returning products to like-new conditions. Remanufactured products are equivalent to the new product in terms of warranty and quality conditions sale (Saavedra et al., 2013). Remanufacturing emerged as an industrial activity after the Second World War. It emerged as a process that resulted in the remanufacturing of battle tanks due to a shortage of raw materials to manufacture new products. Examples in the academic field emerged in the studies developed by Robert Lund in 1984. In the following periods, remanufacturing started to be an area of interest in studies in the field of industry (Hatcher et al., 2011).

The proliferation of remanufacturing in European countries has increased as a result of changes in environmental legislation and the positive change in society's perspective on remanufacturing (Östlin et al., 2008). Specific processes have been established in China to develop specific legislation on reducing the environmental impact of products by remanufacturing and based on sensitivity to the circular economy (Saavedra et al., 2013).

Remanufacturing activities are supported, considering that it will create an important input for the sustainable development of the country and the sustainability of other countries. (Xiang and Ming, 2011).

Advantages and Disadvantages of Remanufacturing

Ayres and others (1997) investigated the impact of firm-level product recovery on economic efficiency and basic input costs. With remanufacturing, product resource input costs were reduced, revealing the firm's real transaction potential and saw a positive change in key performance indicators. It has been revealed that identifying strategies that support the recycling of required products and increasing production potential is an important performance indicator for uncovering new opportunities.

The economic opportunities offered by remanufacturing are competitive price, high quality, customer perception above product value, lowering input costs, high profit return, guaranteed product supply (Sharma et al., 2016). The positive environmental aspects of remanufacturing are expressed in the use of less materials, the safe disposal of hazardous waste, less environmental waste, energy savings and a reduced carbon footprint (Öç, 2013). The social aspects of remanufacturing are expressed in its socially positive image, more jobs and employment, wider market appeal due to lower prices, and new market development opportunities.

Jiang and others (2019) constructed a data-driven remanufacturing system model to evaluate ecological performance. Unlike the traditional qualitative assessment, the quantitative analysis of ecological performance in this model is carried out with an enhanced data analysis. In this model, fourteen criteria such as remanufacturing cost assessment, exhaust emissions, waste disposal, energy consumption, material usage, customer satisfaction, service level, recovery of production due to profit were used.

Ayres and others (1997) suggested that remanufacturing will have a positive effect on the company's expenses as the input costs of parts and components will be lower. In addition, remanufacturing will reduce the amount of waste and create an input for a sustainable environment.

The benefits of the remanufacturing process can be listed as follows (Topoyan, 2005):

- Energy savings
- Creation of new job opportunities as it is a labor-intensive industry.
- Added value for manufactured materials
- Recovery
- Decrease in the price of the product
- Increased competition
- Material recovery
- Solid waste reduction
- Recovery of workforce
- Increase the profitability
- The lower capital requirement
- Recovered value added share is greater than recycling

According to Sharma and others (2016), difficulties in remanufacturing can be listed as quality concerns, customer negative opinions, lack of an established industry for remanufacturing, low customer awareness of remanufactured products, difficulty in selling remanufactured products, difficulty in sourcing products and collecting used products (Sharma et al., 2016).

Other possible difficulties can be listed as follows (Topoyan, 2005):

- Parts recovering a cost
- Difficulty in making forecasts product
- Design changes

Dismantling the remanufactured product is labor intensive. The skill level required will vary depending on the job to be performed. A combination of both skilled and unskilled labor is therefore required. For example, skilled workers are needed to recover the most valuable parts. An even higher level of skill and knowledge is required to distinguish the wear status of various models and valuable components (Ayres et al., 1997).

Remanufacturing in the Automotive Sector

Steinhilp (1998) argues that it is a strong partner in the growth of the automotive industry, considering it is the main driver of the manufacturing industry. Remanufacturing will permanently strengthen the leading role of the automotive industry (Oiko et al., 2011). Automotive parts remanufacturing can be considered an effective strategy to develop and promote sustainability in the automotive industry. Remanufactured automotive spare parts can be supplied at lower prices, contributing to less material and energy use in the process (Zhang et al., 2011).

The National Auto Parts Remanufacturer Association (ANRAP) was founded in 1994. Their aim is to demonstrate the advantages of remanufacturing quality to customers and companies and vehicle maintenance. It also has the reliability, traceability and customer tracking infrastructure of the products produced by the association companies. In addition, since 2011, ANRAP has been organizing workshops for companies and customers in the automotive industry to highlight the industry's benefits and applications of remanufactured quality products, explaining the importance of efficient reverse logistics.

In addition, workshops are held with customers to ask questions about the challenges associated with refurbished and remanufactured products. The results from this are used as source input for the strategies to be created (Saavedra et al., 2013).

Remanufacturing in Turkey Automotive Industry

Gürler (2010) includes remanufacturing activity research in his thesis to examine the situation in Turkey. Enterprises operating in the electrical and electronic industry, automotive industry, tire, engine, metal and machinery industry, printer consumables industry, medical devices industry, and office furniture sectors participated in the study. Businesses think that they can increase their reproduction and profitability with their own technical means. However, it has been concluded that this production process is not yet mature in an environment where customers have a biased approach to remanufactured products and the results of the use of the products remain uncertain.

In a similar study, Gelmez (2013), who determined that the companies producing in the automotive sub-industry do not approach remanufacturing negatively, argues that incentives for remanufacturing should be increased. The author examines the relationship between supply chain performance and business performance and reproduction activities. Arsu and Özdemir (2019) determined the weight of the goal programming model, inventory control, objective function they developed for small and medium-sized businesses as Analytical Hierarchy Process (AHP). In the application, the periodic production, demand, sales, and stock amounts of the three products with the highest amount of remanufacturing of a company that produces automotive spare parts are modeled. When the obtained variable values and deviations from the target were compared with the firm, suggestions were made to increase the efficiency of the enterprise.

Theoretical Framework of Remanufacturing

The Resource-Based View and Remanufacturing Relationship

Traditionally, these resources will not be sustainable unless the strategic management and diversity of resources and internal organizational resources available in a standard industry are not used to analyze an organization's external threats and opportunities. With external pressure and threats, resources will become ineffective (Porter, 1982). According to the resource-based view, strategic resources among firms in an industry may differ, these resources may be mobile, and as a result, differentiation may take time in the long run. Therefore, the resource-based view suggests that a company can achieve economic sustainability by using resources effectively and efficiently (Barney, 1991).

According to Barney, a firm's resources are all the assets, organizational processes, capabilities, firm characteristics, and information controlled by the firm that enable the firm to design and implement strategies that increase its effectiveness and efficiency. Procedures and rules, less formal norms, social communication patterns are also seen as information and resources by organizations. These sources of information include education, social relationships, experience, and the views of employees and managers. For the post-remanufacturing business model, this may mean the ability of the business to learn and improve business processes that will increase efficiency and effectiveness. By building relationships with its suppliers, promoting environmentally and socially responsible practices, and sharing information, the business can develop some of its suppliers' capabilities and resources (Subramoniam et al., 2008).

Such efforts can result in an image of sustainability throughout the supply chain and help improve their economic and ecological business performance. With this perspective, which is in line with Carter and Rogers (2008), social and environmental information resources and results that reinforce economic sustainability propositions when considered as an integrated supply chain may be a more difficult resource to imitate. According to Michaud and Llerena (2006), creating an enviromental sustainable image companies also need to make remanufacturing products that are attractive and acceptable to the end consumer.

Increasing motivation to return used products can be achieved by marketing the quality of remanufactured products. Comparing and testing new and remanufactured products ensures that remanufactured products are as good as new with data. It also saves end users and money due to the lower cost of remanufactured products. This situation constitutes the reason for their preference and a source of motivation for their choice (Subramoniam et al., 2010).

A RESOURCE-BASED VIEW AND REMAN LOGISTICS. Reverse logistics (RL) operations is considerably more complex than just redesing their supply chains. Because of this situation many high-tech manufacturers, logistics operations revise in terms of strategic planning, to third-party logistics provider (3PL) that examines the feasibility of providing outsourcing. Reverse logistics service requirements are analyzed using the RVB approach after reviewing internal resources and capabilities, providing a theoretical basis for determining which reverse logistics operation requirements should be outsourced.

Despite the theoretical emphasis on reverse logistics, development in this area is lacking. RBV is rarely used for the logistics activity of remanufactured automotive parts. There are limited examples of performance and resource allocation from industry stakeholders to various applications such as reverse logistics programs (Xiangru and Wei, 2009). The RBV approach can help the decision-maker as well as the manufacturer identify available resources and capabilities and what third parties need to take action on, distinguish between core competencies and define which activities should be performed in-house or outsourced.

Outsourcing under RBV is a strategic decision for a spare-parts manufacturer that involves outsourcing non-strategic activities deemed necessary for remanufacturing through contracts with more skilled firms to undertake reverse logistics activities to increase competitive advantage (Tajik et al., 2014). Outsourcing of non-strategic activities under RBV is preferred to increase competitive advantages with more skilled companies in reverse logistics activities. It is a strategic decision by the manufacturer about outsourcing non-strategic activities through contractual agreements (Chu et al, 2010).

RBV AND PRODUCTIVITY IN REMANUFACTURING OPERATIONS AND CIRCULAR ECONOMY PERFORMANCE. The digital transformation performance of the supply chain of remanufacturing operations can provide significant opportunities for the circular economy. To develop this transformation, companies need to be closely involved with industry 4.0 and the structuring of their resources to implement leading and core technologies.

Based on the Resource-Based theory, according to research sources examining the effects of industry 4.0, an increase in circular economy performance and reproduction efficiency has been observed after the transformation of operations with industry 4.0 (Bag et al, 2020).

PRODUCTIVITY IN REMANUFACTURING OPERATIONS AND CIRCULAR ECONOMY PERFORMANCE. Remanufacturing is a tool for companies to achieve different goals. There are many ways to increase the efficiency of the business with remanufacturing. Remanufacturing reduces consumption of other raw materials and materials, with proper planning and scheduling, saving labor hours and eliminating monetary losses, and increasing overall equipment efficiency. Hourly capacity increases with reduced cycle times are a great achievement of remanufacturing. In addition, shorter production cycle times compared to their competitors provide such companies with a competitive advantage. remanufacturing; It contributes to the environment and the circular economy and helps to strengthen the economy by focusing on environmental quality elements and increasing resource efficiency and increasing remanufacturing efficiency (Zhang, Liu, et al., 2018).

Remanufacturing is unaffected by uncertainty and risks. It also provides greater visibility to eliminate supply bottlenecks in digital systems. In addition, the availability of market intelligence and data on global supply pricing trends will provide the advantage of controlling supply chain costs in manufacturing operations. It is able to optimize resource inputs and supply cycle times thanks to its high level of visibility. In this way, companies can achieve agile and customer-focused procurement and run a successful remanufacturing operation in this volatile business environment (Bag et al., 2020).

NATURAL-RESOURCE-BASED VIEW AND REMANUFACTURING SUPPLY CHAIN MANAGEMENT. The RBV is restricted to explaining firm-level outcomes and does not consider the environmental impact of firms' activities. In response to this, Hart (1995) developed the NRBV of the firm. In contrast to the RBV, the NRBV respects the constraints of the natural environment and can be seen as a theory of competitive advantage based upon the firm's relationship to the natural environment (Hart, 1995). According to NRBV, one of the sources of competitive advantage is the environment and it wants to examine how this approach produces positive results in the environment. For example, continuous development and lower-cost manufacturing processes can result in lower emissions. With its strategic proactivity feature on the environment, it constitutes an important input in terms of both circular economy and environmental protection (Arag´on-Correa et al., 2008).

With the increasing importance of sustainability, it is seen that the importance of supply chain management that supports the circular economy has come to the fore. In addition, with the increase in sanctions that protect the environment, the content of the supply chain management tools that support the circular economy is changed and updated accordingly. Closed-loop supply chain management is undoubtedly defined as one of the key drivers of sustainability and one of the primary factors in achieving sustainable operations (Kazemi et al., 2018).

NRBV with competitive gains is a harmonious approach to closed-loop supply chain management. Such adaptations are efficiency, cost reduction, waste reduction and reuse of parts after life. System design, process control, and closed-loop design are topics covered by NRBV to explore the value of supply chain management throughout a product's life cycle so that waste parts can be recycled. The circularity identified as the main theme deals with each source's NRBV sources competitively (McDougall, 2018). With the advantages of remanufacturing, competitive advantage can be achieved in the long-term business strategy to tackle environmental challenges. In addition, with remanufacturing, constraints in the ecosystem outside the company can be dynamically managed (Hart, 1995).

NRBV related considerations will be considered in the organization if strategists and organizational theorists in companies generate ideas to reveal the potential of environment-oriented resources and talents for companies to be competitive. To support this, the NRBV has been presented with three interrelated strategies, one of which is the "product management strategy" (Hart, 1995). The "product management concept" seeks to reduce the negative environmental impact of the product life cycle by redesigning existing products or developing new products with lower life cycle costs (Hart, 1995). It has been found that the concept of remanufacturing overlaps with the product management strategy. In this way, it can be seen that it is a more environmentally friendly production method (Ali et al., 2015).

Methodology

Case Study for Creating New Business Model for Aftermarket Spare Parts Product Group

Engines, gearboxes, and turbochargers constitute the most expensive product groups in terms of price in automotive after-sales spare parts and service processes. The need for remanufacturing for these products is increasing day by day in line with customer demands. There is no model in the market where authorized services and authorized renewal centers work together. In this application part, a sustainable and competitive business model in which authorized services and authorized renewal centers can work together has been created separately for each product family.

In this study, the participant observer method was used as the data collection method in the research. Participatory observation is an observation method that allows researchers to obtain information about the activities of the people studied in the natural environment by observing and participating in the activities (DeWalt and DeWalt, 2002). Participant observation is a type of data collection method typically used in qualitative research and ethnography. This type of methodology has been used in different disciplines such as sociology, anthropology, communication studies, human geography, and social psychology.

Its purpose is to try to understand a particular group of individuals and practices by engaging with people in their cultural environment, often over a long period of time. Participant observation has been defined as "learning the daily or routine activities of participants in the researcher environment through analysis or process monitoring" (Kawulich, 2005). For this, area and dealer visits were made and customer insights were collected by the participant of this case. A model was created with agile working methods. In this way, it is aimed to offer quality and guaranteed reman products to customers. There is no other similar study in the after-sales market.

According to Seitz and Peattie (2004), the analysis of the remanufacturing part logistics flow and the good design and management of the flow, the solutions to the problems that may arise, and the design of the flow that will meet the real needs of the most important customer are the problems that can be solved using case studies. Companies that offer sustainable and competitive solutions to customers with Reman products will continue to exist in the market in the long term. In the aftermarket spare parts market where competition is high, the reman solution provides an important competitive advantage for businesses. When evaluated with the Resource

Based View approach, reman spare parts are seen as an important resource input for businesses to continue their processes and assets.

According to Yin (2003), An exploratory case study explores a variety of phenomena characterized by a particular research environment that limits the choice of formulated hypotheses and/or methodology that can be tested. It is intense and is done before larger scale studies are implemented. Discovery situations are appropriate when there is significant uncertainty about the operation of the program, its aims and outcomes (Yin, 2003). Yin also suggests that in a given situation, singular or multiple units of analysis can be studied, thus providing the logic of the second reproduction. A field research case study for four different products, as suggested by Yin, was conducted to evaluate the relevance of the propositions in this article for an automotive reman spare parts center. The case study is used to evaluate strategic factors from the literature review and to clarify questions that require further investigation. A representative case was chosen with four different products; the inspectors had the opportunity to observe and analyze the relevant processes.

According to Stuart and others (2002), a five-stage research model was used in this case study to conduct the case study research proposed. Engine, turbocharger, automatic transmission and injector, which are four different product groups, were evaluated according to this process.

The companies reviewed in this case study are one of the largest automotive remanufacturing spare parts operations centers and dealerships with an independent aftermarket business. Remanufacturing is not carried out in these selected companies. There is a different company remanufacturing operation center for each product group. In this study, a five-stage research model was developed and applied based on Stuart's (2002).

This case study was conducted with twenty-two authorized service points and a chain of remanufactured spare parts revision centers in four different cities in Turkey. The study of Handfield and Melynk (1998) was used as a basis for designing questions directed to customers. It has been questioned why customers with engines, automatic gearboxes, turbos, and injector faults in their vehicle do not prefer authorized service chains. The reasons for the preference of these customers were analyzed. As a result of the analysis of reasons for preference, a business model that would increase the business volume of both the authorized chain services and the authorized revision center has been put forward by producing solutions to the needs of the customers. The business model we developed specifically for this study is based on a flow that incorporates the features specific to Turkish context and the selected sample.

General Research Questions

We aim to provide answers to a certain extend to the general questions that are the major drivers behind this research study:

How does this proposed new circular economy business model contribute to waste reduction and optimization of resource efficiency? How does it contribute to extended product life cycles? In order to find answers to these questions, remanufacturing spare parts in the after-sales spare parts sector were chosen as a sample product for case analysis. With this study, we aim to create a business model for creating value in terms of the circular economy by ensuring that customers prefer these products more. With these preferences of the customers, it is expected that the product groups will be reused, thereby increasing efficiency in operational processes, reducing the amount of waste and increasing the lifespan of the product.

The specific research questions tartgeted at the sample aimed to find out and determine the factors that will enable customers to prefer authorized chain services in engine, turbocharger, automatic gearbox and injector failures.

Data Gathering and Data Analyses

For this study, four product groups were selected. In case of failure of these products, there are alternatives in the after-sales market that do not fully satisfy the customer, do not have a guarantee, and cause different problems. The important point in this study is to create a business model that will enable the customer to prefer the authorized service in case of malfunction and to take action without any problems. For this purpose, insights were collected from customers through questionnaires and site visits were made to authorized chain services. Questions are as follows:

1. Is the customer coming for engine repair?
2. What is your percentage of engine work for total transaction volume?
3. Why do you think customers do not come to the service for engine repair?
4. Do you have easy access to parts for engine repair?
5. Does the service have a contracted remanufacturer to have engine overhaul work done?
6. Does the service have competent personnel to have the engine overhaul work done?

7. Does the service have engine overhaul work and use remaining spare parts?
8. Are there any missing parts about engine rectification?
9. What can be done to increase the engine overhaul business?
10. Does making a bonus system for service employees support the motor business?

Data analyzes were made after the data has been obtained after interviews were done. Comparisons were made with other solutions on the market to further confirm the findings. The product positionings in the market were compared. Differences between existing solutions and customer needs were discussed and detailed questions were asked to the participants in order to understand their real need. Opinions and suggestions have been received. These data were used as input in the proposed developed business model.

Dissemination

The products selected in this case study are the most needed and customer-sensitive products in the engine aftermarket in the automotive aftermarket spare parts industry. The low-priced and low-quality alternative solutions of the products show the need for a business model that offers low-priced, high-quality, and guaranteed solutions to the customer, which was revealed as a result of the work of the authorized services operating in this field. Products are core products associated with the engine product family. The business model that emerged as a result of these products is an adaptable business model for alternators, starters and some other electronic parts.

Current and Proposed Business Models

The current and proposed business model is explained in detail for each product as seen below. In the current situation, each product group is offered to the customer only with an original or sub-industry solution. In the new business model, customers are served with remanufacture spare parts, which provide price advantage and an authorized service guarantee. In this way, it has been observed that the business volume of the services that respond to customer needs has increased.

CURRENT AND PROPOSED BUSINESS MODELS FOR ENGINE PRODUCT FAMILY. For engine product group failures, the option of replacing the engine at the authorized service or renewing the engine as part of

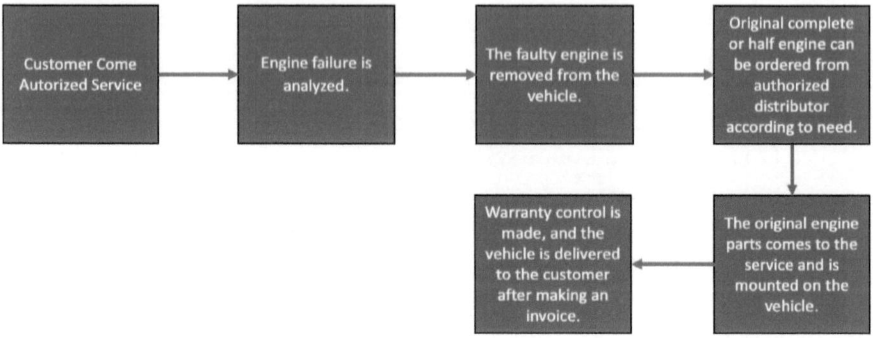

Figure 4.1. Current Business Model
Source: Authors preparation

the original parts is offered to the customer. The business model in these solutions is not preferred over other solutions in the market due to the high-priced customer solution. The current business model is shown in the Figure 4.1. Since the price level remains very high in this business model, it appeals to approximately 2 percent of its customers in the market. Since the customers in the other part are price sensitive, they prefer low-priced and poor-quality alternative solutions in the market instead of authorized services. The research question and purpose of this study is to create a business model at a new price level that will appeal to some of the price sensitive customers of the current business model, and to ensure that some of their customers prefer the authorized service.

NEW BUSINESS MODEL ENGINE REMANUFACTURING PROCESS. After analyzing the current business model, options were evaluated to offer low-priced products to customers. Engine failures that occur when customers come to the service are analyzed and grouped by frequency analysis. This is shown in Figure 4.2 according to failure type. The parts used in these fault groups have been analyzed. The kit was made by grouping the parts that are used in the complete engine revision and that are changed in each revision. When the parts are ordered as kits from an authorized distributor, they are ordered at a more affordable price. This situation seriously reduces the occurrence for the customer, and it is preferable to come to the authorized service. With this kit, the engine remanufacture process solution has emerged as the first and only solution of authorized chain services in the aftersales market. In this respect, it is expected to constitute an input for future studies.

The engine renewal business model created with this kit is the first in its field in terms of sustainability and circular economy. It is expected to

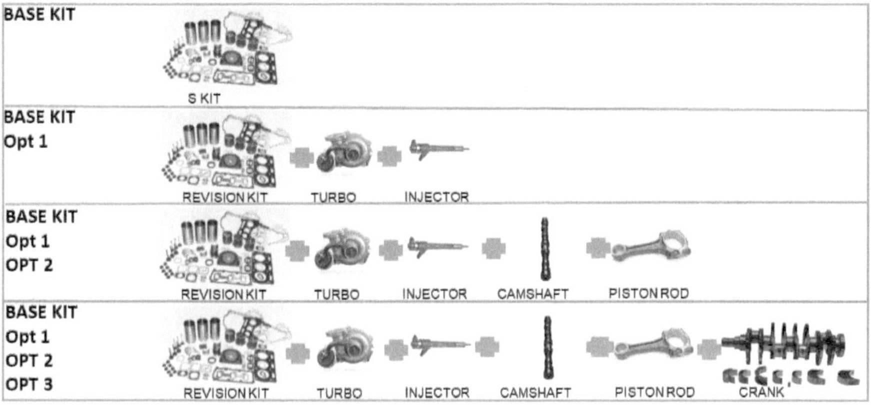

Figure 4.2. Engine failure solutions type
Source: Authors preparation

contribute to different product groups and different sectors in the world literature. The created kits were collected in four main groups. While making this grouping, the types of failures encountered by the customers were taken into consideration. The following picture shows the detailed contents of the kits. The first kit includes only the base revision kit. This includes parts that should definitely be replaced at the lowest level of engine failure. In general, in case of gasket burning, vehicle overheating, or lack of oil, only these parts need to be changed. The second kit is needed in case of malfunctions that are larger than the malfunctions encountered in the first malfunction. In these types of failures, the turbochargers and injectors were damaged along with the engine. While the engine is overhauled, these parts are revised and replaced. In the third kit requirement, if the piston rods are damaged, the engine is overhauled, the turbochargers and injectors are taken to the reman process, and the piston rods are changed. In the fourth case, in addition to the first three groups, the crankshaft is also damaged, and all parts are replaced in the engine shell.

In the new business model, the engine remanufacturing process starts with the order of the relevant kit type from the authorized distributor after the vehicle is diagnosed at the service. After the engine diagnosis is made, it is removed from the vehicle and all the parts that will be changed in the service are removed. The rest is sent to the engine inspection center with the kit. Here the engine is subjected to grinding, honing and similar processes. The engine block is defeated by going through the relevant processes. After the renewal process, the parts to be installed in the renewal center for your kit are attached to the engine block and sent to the service.

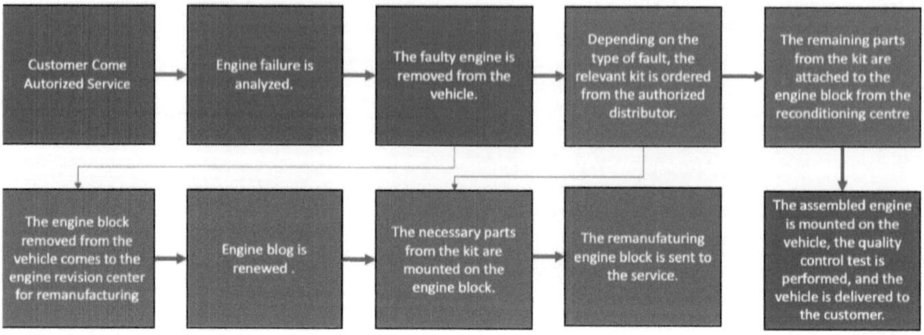

Figure 4.3. New business model
Source: Authors preparation

The service makes the assembly of the engine by attaching the other parts of the kit to the top of the engine block. The reconditioned and reconditioned engine is mounted on the vehicle. After the quality control processes are completed, the vehicle is invoiced and delivered to the customer. The new engine remanufacturing process flow is shown in Figure 4.3.

NEW PROPOSED REMANFACTURE BUSINESS MODELS FOR TURBOCHARGER PRODUCT FAMILY. The damaged turbocharger is removed from the top of the vehicle. The reconditioned turbocharger is available from the supplier (Figure 4.4). The authorized service sends the turbocharger, which it has removed from the top of the vehicle, back to the supplier.

NEW PROPOSED REMANFACTURE BUSINESS MODELS FOR INJECTOR PRODUCT FAMILY. The damaged injector is removed from the top of the vehicle. The reconditioned injector is available from the sup-

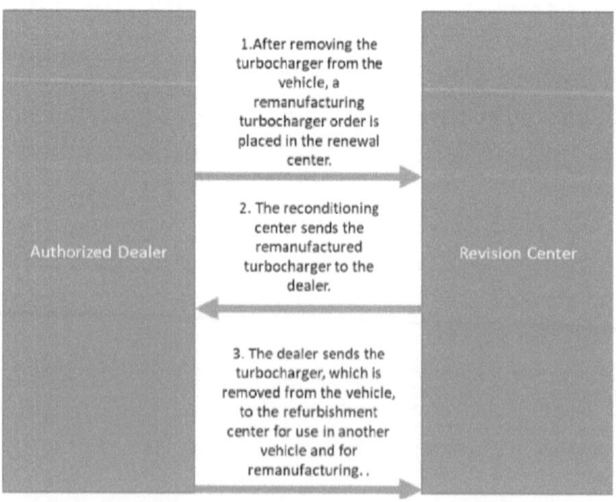

Figure 4.4. New Proposed Remanufacture Business Models for Turbocharger Product Family
Source: Authors preparation

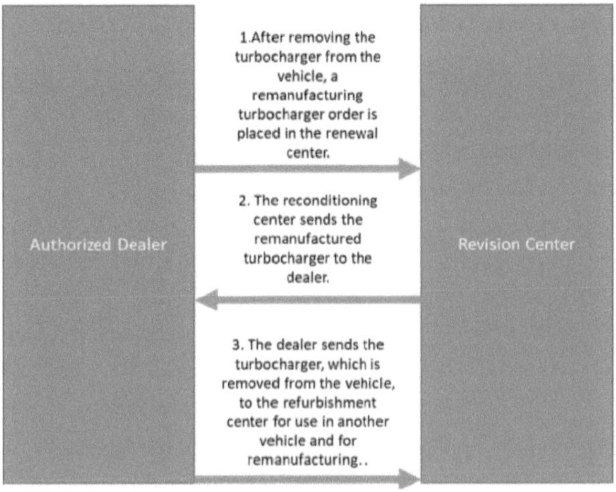

Figure 4.5. New Proposed Remanufacture Business Models for Injector Product Family
Source: Authors preparation

plier. The authorized service sends the injector, which it has removed from the top of the vehicle, back to the supplier (Figure 4.5).

NEW PROPOSED REMANUFACTURE BUSINESS MODELS FOR GEARBOX PRODUCT FAMILY. The damaged automatic gearbox is removed from the top of the vehicle. The reconditioned automatic gearbox is available from the supplier. The authorized service sends the automatic gearbox, which it has removed from the top of the vehicle, back to the supplier (Figure 4.6).

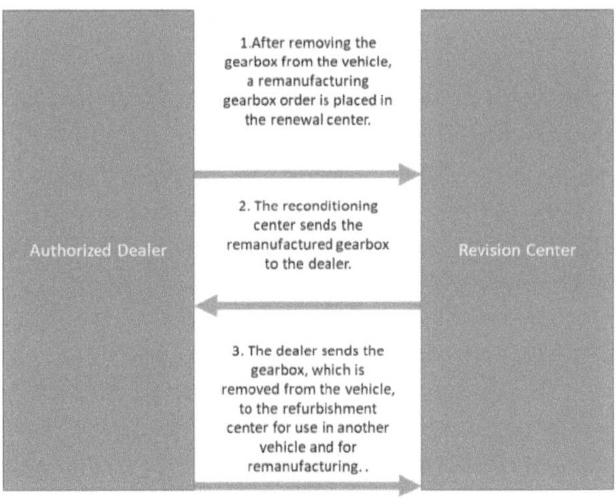

Figure 4.6. New Proposed Remanufacture Business Models for Gearbox Product Family
Source: Authors preparation

Findings and Assessment of Case Study Results

DEVELOPING COOPERATION BETWEEN REVISION CENTERS AND AUTHORIZED SERVICES. In this study, a cooperation agreement was made between the renewal center chain operating in forty-one provinces in Turkey and the authorized services. Within the scope of this cooperation, a price advantage was provided to the customer. In addition, there has been an increase in the business volume of both revision centers and services. Depending on the special agreement, the duration and scope of the warranty given to the customer has been extended. This has been a factor for the customer to prefer the Thanks to this business model and cooperation created between renovation centers and services, it has created the opportunity to adapt it to other product groups. In this way, both the chain services and the renewal center have become viable and competitive authorized service.

LOGISTIC MODEL AND PART SUPPLY AND PRODUCTION SYSTEMS. In the logistics flow created for the engine block renewal process, the dealer sends the engine block, which it has put into the contracted renewal center, together with the relevant kit parts, and takes it back after the renewal process. This flow is actually a logistics flow designed for an outsourced support. The dealer and the renewal center can send the core part for the turbocharger, injector and gearbox product groups and procure the remanufacure part. This flow is provided without loss of time and without creating customer dissatisfaction as in the flow of original spare parts. This logistics flow model will contribute to the literature as an opportunity that can be applied to other sectors and product groups.

REMAN PROCESS STRUCTURE. For the first time in Turkey, the engine renewal process for passenger cars and light commercial vehicles was defined at authorized service centers. With this remanufacturing process, it has been designed to provide benefits in certain areas such as customer satisfaction, increasing the business volume of the dealers and reducing the waste to the environment. It has the feature of being the first and only with the customer mass it addresses and the solution it offers.

A GENERALIZABLE REMANUFACTURING SOLUTION. The study was carried out in Turkey and for a certain vehicle group customer. Considering the positive effects of the study on the circular economy and the environment, the business model can be applied to different sectors and generalized.

For example, this business model created for passenger and light commercial vehicle models can be designed as a generalizable business model for the renewal of electric motors, which are needed in the digitalizing world.

POZITIVE IMPACT FOR ENVIRONMENT. With this study, it is aimed to eliminate the negative effects on the environment with the remanufacturing process of some parts in the engine product group, especially in the automotive sector, which constitutes the majority of scrap waste. It is expected that the amount of metal waste generated as a result of this study will decrease. When evaluated in terms of sustainability and resource-based approach, the reman process is of serious importance for spare part manufacturers and consumers.

Conclusion

A comprehensive review of the existing literature in the fields of remanufacturing and the automotive aftermarket and an analysis of the automotive aftermarket independent aftermarket industry have been conducted to identify key actors. A needs analysis was conducted to develop a new business model for remanufacturing in this area. The model was obtained and tested as a result of a case study. With the case study, points that were not taken into account while developing the business model were discovered. In this study, the focus was on the automotive independent aftermarket and authorized services. The actors examined in the study need to adopt the sustainability and circular economy approach by applying remanufacturing to automotive spare parts.

While responding to customer needs by prioritizing sustainable growth and competitiveness, authorized services should pay attention to issues such as product life cycle perspective, less raw material extraction and consumption, as well as reduced energy consumption. Therefore, having the competence to use remanufactured parts emerges as the only solution for businesses. With remanufacturing, designing and commissioning a business model in which parts of the products or all products can be renewed after use emerges as the optimum solution. From this point of view, the need to model a cyclical material flow for supply chain management also arises.

Automotive aftermarket remanufacturing applications emerged as a strategic requirement for the actors in the value chain at the end of the study. Authorized services have reached the competence to offer remanufactured parts to their customers with the developed model at the end of the study. In this way, they will be able to maintain their presence in the market in a sustainable and competitive way. With this study, it has made a very important contribution to the literature by creating a business model that enables the use of remanufactured parts, especially in highly competitive product groups, for the first time in Turkey.

To sum up, the most important result of this study has been the increase in the business volumes of authorized services by making the engine remanufacturing process, which had higher costs and was not preferred by the customers before, preferable in authorized services. The new business model shown in Figure 4.3 has been developed to address the missing points of the new business model shown in Figure 4.1 and to meet customer needs. With the newly created business model as shown in Figure 4.3:

- Transaction cost has been reduced.
- Customer satisfaction has increased.
- The rate of preference of authorized services by customers has increased.
- The use of remanufactured parts has increased, and it has become a contribution to the environment and circular economy and to the country's economy.
- Contribution was made to employment.
- As the business volume of authorized services increased, profit margins increased.

In future studies, the business model can be generalized for different product groups. The business model can be made usable by authorized services by making modifications to other needs in the aftersales market, including the remanufacturing method. In this way, the assets of the relevant actors can be sustained in a sustainable and competitive way.

Depending on these developments, it is inevitable that supply chain models that support the circular economy and are less harmful to the environment will emerge in the future. With the results obtained from the study, it is expected that new remanufacturing processes will emerge in the short- and long-term futures to help authorized services better achieve their goals and objectives.

References

Ali, F., Sriram, P. K., Alfnes, E., Brett, P. O., and Fet, A. M. (2015). Remanufacturing as a sustainable strategy in Shipbuilding Industry. In *IFIP International Conference on Advances in Production Management Systems* (pp. 232–39). Springer, Cham.

Aragón-Correa, J. A., Hurtado-Torres, N., Sharma, S., and García-Morales, V. J. (2008). Environmental strategy and performance in small firms: A resource-based perspective. *Journal of environmental management, 86*(1), 88–103.

Armacost, R. L., Balakrishnan, D., and Pet-Armacost, J. (2002). Design for remanufacturability using QFD. In *IIE Annual Conference. Proceedings* (p. 1). Institute of Industrial and Systems Engineers (IISE).

Arsu, T., and Özdemir, A. (2019). Hedef Programlama ve Analitik Hiyerarşi Prosesi (AHP) ile Yeniden Üretim Sistemlerinin Stok Kontrolünün İncelenmesi. *Afyon Kocatepe Üniversitesi Sosyal Bilimler Dergisi, 21*(4), 1230–45.

Ayres, R., Ferrer, G., and Van Leynseele, T. (1997). Eco-efficiency, asset recovery and remanufacturing. *European Management Journal, 15*(5), 557–74.

Bag, S., Dhamija, P., Gupta, S., and Sivarajah, U. (2020). Examining the role of procurement 4.0 towards remanufacturing operations and circular economy. *Production Planning & Control*, 1–16.

Barney, J. (1991). Firm resources and sustained competitive advantage. *Journal of management, 17*(1), 99–120.

Carter, C. R., and Rogers, D. S. (2008). A framework of sustainable supply chain management: moving toward new theory. *International journal of physical distribution & logistics management.*

Chu, D. H., Guo, H. Y., and Zhang, H. (2010). A resource-based view approach for reverse logistics service. In *2010 Seventh International Conference on Fuzzy Systems and Knowledge Discovery* (Vol. 6, pp. 2716–20). IEEE.

Ferrer, G., and Whybark, D. C. (2001). Material planning for a remanufacturing facility. *Production and Operations Management, 10*(2), 112–24.

Gelmez, E. (2013). Ters tedarik zinciri yönetimi bağlamında yeniden üretim uygulamaları: Konya otomotiv yan sanayi örneği, (Yayımlanmamış yüksek lisans tezi). *Selçuk Üniversitesi, Konya.*

Giutini, R., and Gaudette, K. (2003). Remanufacturing: The next great opportunity for boosting US productivity. *Business Horizons, 46*(6), 41–48.

Gray, C., and Charter, M. (2007). Remanufacturing and product design: designing for the 7th generation.

Güner, H., and Güner, A. R. (2020). Sürdürülebilir Bir Çevre İçin Yeniden Üretim: Bir Literatür İncelemesi. *European Journal of Engineering and Applied Sciences, 3*(2), 18–26.

Gungor, A., and Gupta, S. M. (1999). Issues in environmentally conscious manufacturing and product recovery: a survey. *Computers & Industrial Engineering, 36*(4), 811–53.

Gürler, İ. (2010). *Yeniden üretim sürecinde tasarım, planlama, lojistik faaliyetlerinin incelenmesi ve Türkiye açısından değerlendirilmesi* (Doctoral dissertation, DEÜ Sosyal Bilimleri Enstitüsü).

Handfield, R. B., and Melnyk, S. A. (1998). The scientific theory-building process: a primer using the case of TQM. *Journal of operations management, 16*(4), 321–39.

Hart, S. L. (1995). A natural-resource-based view of the firm. *Academy of management review, 20*(4), 986–1014.

Hatcher, G. D., Ijomah, W. L., and Windmill, J. F. C. (2011). Design for remanufacture: a literature review and future research needs. *Journal of Cleaner Production, 19*(17–18), 2004–14.

Ijomah, W. L., McMahon, C. A., Hammond, G. P., and Newman, S. T. (2007). Development of design for remanufacturing guidelines to support sustainable manufacturing. *Robotics and Computer-Integrated Manufacturing, 23*(6), 712–19.

Ijomah, W. L., McMahon, C. A., Hammond, G. P., and Newman, S. T. (2007). Development of design for remanufacturing guidelines to support sustainable manufacturing. *Robotics and Computer-Integrated Manufacturing, 23*(6), 712–19.

Jiang, Z., Ding, Z., Zhang, H., Cai, W., and Liu, Y. (2019). Data-driven ecological performance evaluation for remanufacturing process. *Energy Conversion and Management, 198*, 111–844.

Kawulich, B. B. (2005, May). Participant observation as a data collection method. In *Forum qualitative sozialforschung/forum: Qualitative social research* (Vol. 6, No. 2).

Kazemi, N., Modak, N. M., and Govindan, K. (2019). A review of reverse logistics and closed loop supply chain management studies published in IJPR: a bibliometric and content analysis. *International Journal of Production Research, 57*(15–16), 4937–60.

Ljunglöf, J., and Nisser, H. (2017). KPIs in a service organization-a case study of Axfood IT.

McDougall, N. (2018, November). A natural-resource-based approach to closed-loop supply chain management: competitive environmental sustainability in the fashion sector. In *6th International Workshop on Luxury Retail, Operations and Supply Chain Management.*

Michaud, C., and Llerena, D. (2006). An economic perspective on remanufactured products: industrial and consumption challenges for life cycle engineering. *Proceedings of LCE2006*, 543–48.

Musante DeWalt, K., and DeWalt, B. R. (2002). Participant observation: A guide for fieldworkers. *Walnut Creek.*

Nasr, N., and Thurston, M. (2006). Remanufacturing: A key enabler to sustainable product systems. *Rochester Institute of Technology,*

Öç, B. (2013). *Sürdürülebilir tasarım: Ürün tasarımı ve üretimi temelinde malzemelerin geri dönüştürülmesi bilinci* (Doctoral dissertation, Fen Bilimleri Enstitüsü).

Oiko, O. T., Barquet, A. P. B., and Ometto, A. R. (2011). Business issues in remanufacturing: two Brazilian cases in the automotive industry. In *Glocalized Solutions for Sustainability in Manufacturing* (pp. 470–75). Springer, Berlin, Heidelberg.

Östlin, J., Sundin, E., and Björkman, M. (2008). Importance of closed-loop supply chain relationships for product remanufacturing. *International Journal of Production Economics, 115*(2), 336–48.

Porter, M. E. (1982). Competitive strategy. *RAE-Revista de Administração de Empresas*, *22*(2), 44–46.

Rose, C. M. (2001). *Design for environment: a method for formulating product end-of-life strategies*. Stanford University.

Saavedra, Y. M., Barquet, A. P., Rozenfeld, H., Forcellini, F. A., and Ometto, A. R. (2013). Remanufacturing in Brazil: case studies on the automotive sector. *Journal of Cleaner Production*, *53*, 267–76.

Seitz, M. A. (2007). A critical assessment of motives for product recovery: the case of engine remanufacturing. *Journal of Cleaner Production*, *15*(11–12), 1147–57.

Seitz, M. A., and Peattie, K. (2004). Meeting the closed-loop challenge: the case of remanufacturing. *California management review*, *46*(2), 74–89.

Sharma, V., Garg, S. K., and Sharma, P. B. (2016). Identification of major drivers and roadblocks for remanufacturing in India. *Journal of cleaner production*, *112*, 1882–92.

Steinhilper, R. (1998). Remanufacturing-The ultimate form of recycling. *Fraunhofer IRB Verlag*.

Steinhilper, R. (2001, December). Recent trends and benefits of remanufacturing: from closed loop businesses to synergetic networks. In *Proceedings Second International Symposium on Environmentally Conscious Design and Inverse Manufacturing* (pp. 481–88). IEEE.

Stuart, I., McCutcheon, D., Handfield, R., McLachlin, R., and Samson, D. (2002). Effective case research in operations management: a process perspective. *Journal of operations management*, *20*(5), 419–33.

Subramoniam, R., Huisingh, D., and Chinnam, R. B. (2009). Remanufacturing for the automotive aftermarket-strategic factors: literature review and future research needs. *Journal of Cleaner Production*, *17*(13), 1163–74.

Subramoniam, R., Huisingh, D., and Chinnam, R. B. (2010). Aftermarket remanufacturing strategic planning decision-making framework: theory & practice. *Journal of Cleaner Production*, *18*(16–17), 1575–86.

Subramoniam, R., Subramoniam, S., Huisingh, D., and Krishnan Kutty, K. V. (2008). Mass customization ea key driver for the emerging automotive aftermarket business model. *International Journal of Global Business*, *1*(1), 1–25.

Sundin, E. (2004). *Product and process design for successful remanufacturing* (Doctoral dissertation, Linköping University Electronic Press).

Tajik, G., Azadnia, A. H., Ma'aram, A. B., and Hassan, S. A. H. (2014). A hybrid fuzzy MCDM approach for sustainable third-party reverse logistics provider selection. In *Advanced Materials Research* (Vol. 845, pp. 521–26). Trans Tech Publications Ltd.

Thierry, M., Salomon, M., Van Nunen, J., and Van Wassenhove, L. (1995). Strategic issues in product recovery management. *California management review*, *37*(2), 114–36.

Topoyan, M. (2005). Yeniden üretim sistemleri için sürdürülebilir ürün tasarımlarının oluşturulması.

Xiang, W., and Ming, C. (2011). Implementing extended producer responsibility: vehicle remanufacturing in China. *Journal of Cleaner Production, 19*(6–7), 680–86.

Xiangru, M., and Wei, S. (2009, July). Network design based on closed-loop reverse logistics recycling. In *2009 IITA International Conference on Control, Automation and Systems Engineering (case 2009)* (pp. 533–36). IEEE.

Yin, R. K. (2003). Case study research. Desing and methods. *Appl. Soc. Res. Methods Ser, 5.*

Yusop, N. M., Wahab, D. A., and Saibani, N. (2016). Realising the automotive remanufacturing roadmap in Malaysia: challenges and the way forward. *Journal of Cleaner Production, 112*, 1910–19.

Zhang, T., Chu, J., Wang, X., Liu, X., and Cui, P. (2011). Development pattern and enhancing system of automotive components remanufacturing industry in China. *Resources, Conservation and Recycling, 55*(6), 613–22.

Zhang, Y., Liu, S., Liu, Y., Yang, H., Li, M., Huisingh, D., and Wang, L. (2018). The 'Internet of Things' enabled real-time scheduling for remanufacturing of automobile engines. *Journal of cleaner production, 185*, 562–75.

Antecedents of Consumers' E-Waste Recycling Behavioral Intentions Based on Theory of Planned Behavior

5

TÜLIN URAL AND OĞUZ OYPAN

Introduction

Various factors such as urbanization, rapid developments in technology, and population growth bring about an increase in product variety, more consumption, and shortened product life cycles. As a result of these cases, solid waste rates increase significantly. Especially in recent years, with the rapid growth of the global consumer electronics market, e-waste has become one of the biggest solid waste problems in the modern world. In 2019, there was a total of 53.6 Mt of e-waste worldwide. It is also estimated that there will be 74.7 Mt of e-waste worldwide in 2030 (Forti et al., 2020).

E-waste contains both harmful and precious materials. Harmful elements such as chromium, mercury, cadmium, and lead in e-waste by polluting the soil and water seriously affect human health, natural resources and the environment. Besides, e-waste contains many precious metals such as iron, copper, aluminum, gold, and silver. For this reason, recycling e-waste is of great importance in terms of economy and environment. However, the recycling rates of e-waste are at very low levels. For instance, in 2019, only 17.4 pecent of e-waste could be recycled. One of the main reasons why e-waste recycling rates are so low is that consumers are reluctant to e-waste recycle (Kumar, 2019). Therefore, it is of great importance to determine the antecedents that motivate consumers in the recycling of e-waste. Past research results show that theory-based research is needed to understand consumers' e-waste recycling intention. In this research, we aim to contribute to this gap.

The study is examined the consumers' intention toward recycling e-waste based on the theory of planned behavior (TPB). TPB is one of the most preferred theories in researching environmental behaviors. Many researchers have preferred the TPB model to examine various environmental behaviors, such as recycling behavior (Botetzagias et.al., 2015), water-saving behavior (Lowe et.al., 2015), green product purchasing behavior (Chen and Tung, 2016), energy-saving products purchasing behavior (Nguyen et.al, 2016), and green hotel accommodation intention (Verma and Chandra, 2018).

However, many researchers emphasized that TPB should be increased by adding different variables to increase its explanatory power. (Perugini and Bagozzi, 2001; Chen et al., 2016; De Leeuw et al., 2015; Ajzen, 2020). Based on these recommendations, the current study included the warm glow variable in the original model and investigated how this variable influences the intention to recycle e-waste. The warm glow concept refers to the psychological satisfaction that people feel as a result of their environmental behavior (Taufik et al., 2014). In previous studies, environmental behaviors were examined based on the warm glow variable (Nguyen et al., 2017; Hartmann et al., 2017). However, to the best of the authors' knowledge, e-waste recycling behaviors based on the warm glow variable have not been investigated. In previous studies, e-waste recycling intention was investigated based on rational variables such as knowledge (Arain et al., 2020), convenience (Zhang et al., 2019), awareness (Kumar, 2019), risk perception (Dhir et al., 2021), economic motivation (Wang et al., 2019). However, there is a gap based on intrinsic motivation variables. In this respect, this study makes a magnificent contribution to environmentally friendly behavior research.

Conceptual Framework and Hypotheses Development

Theory of Planned Behavior

TPB (Ajzen 1991), an extended model of the Reasoned Action Theory (TRA), is one of the most popular theories used to explain human behavior. According to TPB, behavioral intentions are the main determinant of actual behavior. Behavioral intention indicates that an individual is cognitively ready to take action to perform a certain behavior (Ajzen, 1991). According to the theory, if an individual has an intention to perform a certain behavior, the probability of performing that actual behavior will increase. According to the theory, three concepts are effective in the formation of intention (Ajzen, 1991). These are attitude, subjective norm, and perceived behavioral control (PBC).

Attitude refers to an individual's evaluation (positive/negative) of performing a certain behavior (Ajzen, 1991). Having a positive attitude towards the behavior will lead to the formation of positive intentions towards that behavior. When the current literature is examined, many studies have revealed that attitudes are one of the strongest antecedents that affect environmental behaviors (Han and Yoon, 2015; Yadav and Pathak, 2016; Yu and Lee, 2019; Si et.al., 2020). On the other hand, many researchers have examined how attitudes affect the intention toward recycling e-waste. For instance, Aboelmaged (2021) revealed that attitudes towards e-waste recycling behavior have a significant effect on young adults' intention toward recycling e-waste. Similarly, Wang and others (2018) determined that attitudes behavior significantly affect behavioral intention toward recycling e-waste. Based on these arguments, the following hypothesis was developed:

H1: Attitude towards recycling significantly and positively affect the behavioral intention towards recycling e-waste.

Subjective norms are social pressures that shape an individual's behavior. The individual either performs or does not perform a behavior through this social pressure (Ajzen, 1991). Individuals are more likely to perform behaviors that meet the expectations of the social environment (family, relatives, friends, etc.). In many studies such as sustainable transportation preference (Cai et al., 2019), green hotel stay intention (Verma and Chandra, 2018), energy-saving behavior (Nie et al., 2019), organic food consumption behavior (Scalco et al., 2017), it has been revealed that subjective norm has an important effect on behavioral intentions. Wang and others (2019) determined that subjective norms significantly influence households' intention to recycle e-waste online. Based on these arguments, the following hypothesis was developed:

H2: Subjective norms significantly and positively affect the behavioral intention towards recycling e-waste.

PBC refers to the "perceived ease or difficulty of performing a particular behavior" (Ajzen, 1991, p. 188). The greater the perception of control that an individual can cope with the difficulties of performing a behavior, her/his intention to perform this behavior will be high (Ajzen, 1991). Perception of control is determined by internal factors (the person's skills and abilities) and external factors (time, money, and availability). Previous studies have revealed that PBC has an important role in explaining environmental behavior (Judge et al., 2019). For instance, Nimri and others (2020) found that PBC significantly influences green hotel stay intention. Chen and Tung (2014) revealed that PBC positively influences the intention to stay in a green hotel. Based on these arguments, the following hypothesis was developed:

H3: Perceived behavior control significantly and positively affects the behavioral intention towards recycling e-waste.

Extension of the TPB

Many researchers have explained environmental behavior by adding different variables to TPB. For instance, Botetzagias and others (2015) investigated the recycling behavior of consumers by adding the moral norms variable. Alzubaidi and others (2020) examined the factors affecting consumers' behavior toward environment protection by adding the variables of materialism and innovativeness. In this study, the original TPB model was expanded based on the warm glow variable, as shown in Figure 5.1.

The concept of warm glow expresses the psychological satisfaction that pro-social behavior creates in the person (Isen, 1970; Andreoni, 1990). The warm glow is one of the important intrinsic motivation factors used to explain environmental behaviors. Consumers who choose environmental behavior experience the warm glow as a result of the psychological of contributing to the environment. Individuals with a high sense of warm glow are more intended to engage in environmental behaviors (Hartman and Apaolaza-Ibanez, 2021). Many studies have found results that support this view. For instance, Hartmann and others (2017) stated that consumers with a high sense of warm glow prefer sustainable energy products with higher prices than traditional energy products. Nguyen and others (2017) revealed that the warm glow effect increases participation in recycling behavior as a result of their studies. Therefore, we propose the following hypothesis:

H4: Warm glow significantly and positively affects the behavioral intention towards recycling e-waste.

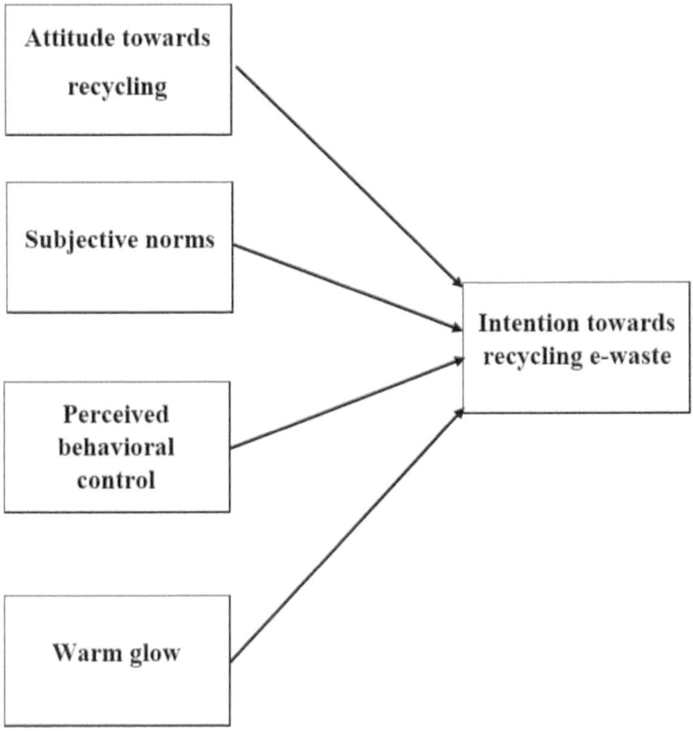

Figure 5.1. Conceptual Model
Source: Authors preparation

Methodology

Sampling Procedure and Measurement

To test the theoretical relationships, the survey was carried out on consumers. The population of the study consists of consumers who are over the age of eighteen living in Turkey. Since it is not possible to reach all of the consumers that make up the main mass, non-probability sampling is used and the convenience-sampling method is preferred for collecting data. Afterward, an online questionnaire was used for data collecting in the period between August 2021 and September 2021. Data were obtained from social media platforms using an online questionnaire created via Google form. The sample consists of 218 consumers. First, a pilot study with thirty consumers has been operated. There was no problem related to understanding the survey questions. Validated instruments in existing literature were adopted in this study. A three-item scale has been adapted from

Wang and others (2016) to measure the attitude towards the recycling construct. A three-item scale measuring the subjective norms construct was adapted from Delcea and others (2020). A three-item scale was adapted from Aboelmagd, M. (2021) for measuring the PBC construct, a four-item scale was adapted from Nguyen et al. (2017) to measure the warm glow construct, and a three-item scale was adapted from Kumar (2019) for measuring the intention behavior construct. All items of scales were also adapted for e-waste recycling of the electric and electronic equipment by researchers. To provide validity, the questionnaire was first officially translated from English to Turkish and later, translated from Turkish back to English. The constructs were measured by a five-point Likert scale ranging from "strongly disagree" and "strongly agree" (see Table 5.1).

In structural equation modeling (SEM), confirmatory factor analysis (CFA) and path analysis has been performed on the pooled data simultaneously by using the AMOS 25. software package.

Analyzes and Findings

Some key features of the sample are summarized as follows. Approximately 65.1 percent of the 218 participants are male. The majority of the participants are high school (50.9 percent) and university graduates (30.3 percent). Nearly 26.1 percent of participants have a monthly income between 5,001 and 7,000 Turkish Lira. Of the total participants, 46.3 percent are 25–34 years old.

Confirmatory factor analysis has been applied to examine the uni-dimensionality, reliability, and validity of the measurement model. Path analysis has been used to test the relationships between proposed constructs which are the "attitude towards recycling," "subjective norms," "PBC," "warm glow," and "intention of e-waste recycling." The findings of the study show that all factor loadings of the measurement model items ranged from 0.62 to 0.93 and are significant (Table 5.1). Most studies indicate that factor loadings must be larger than 0, 5 to reach better results (Truong and McColl, 2011; Hulland, 1999). We did not need to remove any items from the model. Therefore, the uni-dimensionality requirement was achieved.

The reliability of latent variables has been tested by internal reliability, composite reliability (CR), and average variance extracted (AVE) values, The fact that the Cronbach alpha value of each scale is above a certain threshold indicates that this scale has internal reliability. The Cronbach's alpha, CR, and AVE findings are presented in Table 5.2. It is observed that

Table 5.1. Items' Factor Loadings

Code/Variables	*Std. Factor Loadings*	*P Values*
"Attitude towards recycling"		
ar1: "E-waste recycling makes me feel very satisfied."	0.93	***
ar2: "E-waste recycling contributes to the society."	0.89	***
ar3: "E-waste recycling is everyone's responsibility."	0.66	***
"Subjective norms"		
sn1: "My family expects me to engage in e-waste recycling behavior."	0.83	***
sn2: "My friends expect me to engage in e-waste recycling behavior."	0.90	***
sn3: "Other persons with whom I am in contact expect me to engage in e-waste recycling behavior."	0.84	***
"Perceived behavioral control"		
pbc1: "I know what electronic waste can be recycled."	0.77	***
pbc2:" I know how to recycle my electronic waste."	0.83	***
pbc3: "I can succeed in recycling electronic waste."	0.62	***
"Warm glow"		
wg1: "I like to contribute to the protection of the environment by recycling."	0.81	***
wg2: "I have the feeling of contribution to the well-being of humanity when recycling."	0.85	***
wg3: "Recycling household wastes makes me feel good."	0.93	***
wg4: "I am happy with myself whenever recycling household wastes."	0.83	***
"Intention behavior"		
in1: "I intend to make plans to recycle E-waste regularly."	0.79	***
in2: "I intend to recycle E-waste."	0.87	***
in3:" I intend to put extra effort into recycling E-waste."	0.86	***

***: p value < 0.01

Source: Authors calculation

Cronbach's alpha values ranged between 0.78–0.91. Further, CR values are found as greater than the 0.70 thresholds (Hair et al., 2016). Values are between 0.79 and 0.98. Lastly, AVE results were higher than the 0.50 threshold. They ranged from 0.55 to 0.74. These findings are evidence of the existence of construct reliability of the measurement model (Hair et al., 2012). Convergent validity has been tested by factor loadings and AVE. All the factor loadings of the measurement model items have been found significant (see Table 5.1). Thus, taking into account the factor loadings together with AVE values, it can be seen that convergent validity exists because the results have been found required level (Hair et al., 2012).

Table 5.2. Construct Reliability

	Cronbach's Alpha (Above 0,70)	*Composite Reliability (Above 0,60)*	*AVE (Above 0,50)*
"Attitude towards recycling"	0.85	0.98	0.70
"Subjective norms"	0.89	0.89	0.74
"Perceived behavior control"	0.78	0.79	0.55
"Warm glow"	0.91	0.92	0.73
"Intention of e-waste recycle"	0.87		

Source: Authors calculations

The absence of unnecessary items in the measurement model is an indicator of discriminant validity. When the modification indices were checked, it was determined that there were no unnecessary measurement items. In addition, the correlation coefficients between exogenous variables were found to be less than 0.85. The results revealed that the discriminant validity of the model is accomplished. Moreover, the Fornell-Larcker criterion (1981) was applied to test discriminant validity. According to this criterion, the square root of AVE values of all the reflective constructs should be higher than the inter-construct correlations. This condition states the discriminant validity. In the model, findings state that the discriminant validity is acceptable. In Table 5.3, bold diagonal values show the squared root of AVE values, and off-diagonal values represent the correlations of each construct with other constructs.

Based on all of these figures, the measurement model can be accepted (Hu and Bentler, 1999). Goodness of fit values were found as $\chi 2$ / df = 2.340; RMESA = 0.07; NFI = 0.91; CFI = 0.95; IFI = 0.95; TLI = 0.94. The fitness indexes are at the desired level. Construct validity is provided, and the structural model is fit. These figures show that it is acceptable as it falls within the range of values determined by Hu and Bentler (1999).

The structural model was tested by the Maximum Likelihood estimation method, and the bootstrap samples were stated as 5000. R^2 values of the latent endogenous variable were found as 0.58 for the structure of the intention of e-waste recycling. This indicates that 58 percent of

Table 5.3. Discriminant Validity-Fornell*Larcker Criterion

	"Attitude towards recycling"	*"Subjective norms"*	*"Perceived behavioral control"*	*"Warm glow"*
"Attitude towards recycling"	**0.84**			
"Subjective norms"	0.55	**0.86**		
"Perceived behavioral control"	0.19	0.53	**0.74**	
Warm glow	0.73	0.69	0.32	**0.85**

Source: Authors calculations

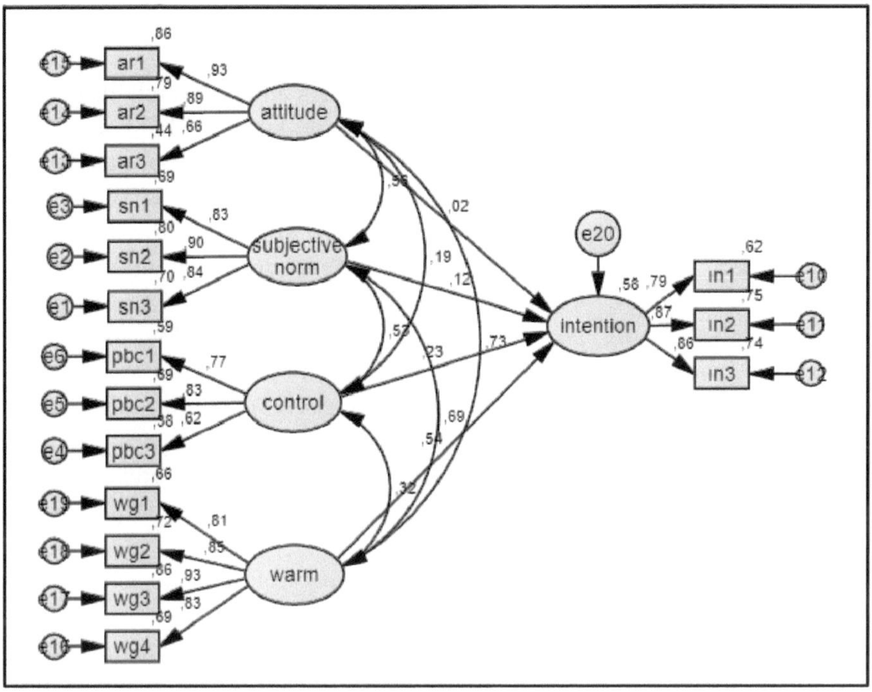

Figure 5.2. Full SEM results
Source: Authors preparation

the intention of e-waste recycling could be estimated by four exogenous constructs, namely, attitude towards recycling, subjective norms, PBC, and warm glow. As shown in Figure 5.2, the results of path analysis indicate that the PBC (β = *0.23, p < 0.01*) and warm glow (β = *0.54, p < 0.01*) variables positively and significantly affect the intention of e-waste recycling. However, the coefficient of the path from the attitude towards recycling to the intention of e-waste recycle (β = *0.02, p = 0.78*), and the coefficient of the path from social norms to the intention of e-waste recycle (β = *0.12, p = 0.21*) was found statistically insignificant (Table 5.4 and Figure 5.2)). Consequently, while the H_3, H_4 hypotheses are supported, the H_1, H_2 hypotheses are rejected.

Table 5.4. Standardized Path Coefficients

PATHS	*Std. Conf.* β	*Std. Error*	*P-value*	*Significant*
recycling	0.02	0.107	0.78	Insignificant
recycling	0.12	0.084	0.21	Insignificant
recycling	0.23	0.082	0.00**	Significant
recycling	0.54	0.108	0.00**	Significant

Source: Authors calculations

Results, Discussions, and Suggestions

Our study aims to examine the antecedents affecting consumers' e-waste recycling intention. Therefore, we developed a new research model by adding the warm glow variable to the original model of the PDT, based on a literature survey. The findings show that the PBC variable positively and significantly affects the intention of e-waste recycling. That is, the greater the perception of control that an individual can cope with the difficulties of performing recycling of e-waste behavior, the more intention it is to perform this behavior. Wang and others (2018) determined as a result of their research that PBC positively affects the e-waste recycling intention.

Further, the results reveal that warm glow is one of the key precursors of e-waste recycling intention. In this respect, consumers who obtain psychological satisfaction as a result of environmentalist behavior are more likely to recycle e-waste. This finding supports previous literature that warm glow positively affects consumers' intention of environmental behavior (Bhutto et al., 2021).

However, it's found that insignificant impact of subjective norms on the intention of e-waste recycling. Although it's surprising, it can be interpreted on basis of current literature and our sample's characteristics. Previous studies have confirmed that subjective norms do not affect consumers' environmental behavioral intentions (Tan et al., 2015). Also, Gao and others (2017) argued that more educated and younger individuals are more rational and have their ideas. We can say that our second hypothesis is not supported because most of the participants in this study were educated and young. Another surprising result is that attitudes toward recycling do not affect the intention of e-waste recycling. Lack of trust, lack of environmental awareness, and lack of social awareness may have caused the attitude-intention gap (Elhaffar et al., 2020).

In the framework of these results, some policy recommendations can be made: First, e-waste recycling facilities and infrastructures should be improved and integrated with internet technologies to increase the perception of control over the behavior of individuals. In addition, the level of knowledge of individuals on how to recycle e-waste should be increased through various advertising and promotion channels. Because, as the level of knowledge about a behavior increases, the individual's perception of control over that behavior will also increase. Second, in communication campaigns for e-waste recycling, messages and themes should be used to increase the intrinsic satisfaction that recycling will create in the heart of a person.

The present study includes suggestions for future research and several limitations. Researchers working in the green marketing field may pay attention to the importance of the intrinsic motivation factors to direct consumers to recycle the waste of electric or electronic equipment. This study includes theory-driven arguments and cross-sectional data in testing hypotheses. To evaluate the intrinsic motivation factors better, there is a need for experimental studies. The convenience sampling method is used in the analysis. It can be stated that this sample group may not represent the general population. It is recommended that future studies should be carried out with a larger sample of respondents to get an overall picture of the consumers' intention towards recycling e-waste. This provides to enhance the generalizability of the findings.

Briefly, this study will captivate researchers and practitioners due to demonstrating the PBC and warm glow feelings positively and significantly influence the intention of e-waste recycling.

References

Aboelmaged, M. (2021). E-waste recycling behavior: An integration of recycling habits into the theory of planned behavior. *Journal of Cleaner Production, 278*, 124–82.

Ajzen, I. (1991). The theory of planned behavior. *Organizational Behavior and Human Decision Processes*, 50(2), 179–211.

Ajzen, I. (2020). The theory of planned behavior: Frequently asked questions. *Human Behavior and Emerging Technologies*, *2*(4), 314–24.

Alzubaidi, H., Slade, E. L. and Dwivedi, Y. K. (2021). Examining antecedents of consumers' pro-environmental behaviors: TPB extended with materialism and innovativeness. *Journal of Business Research*, *122*, 685–99.

Andreoni, J. (1990). Impure altruism and donations to public goods: A theory of warm-glow giving. *The Economic Journal*, *100*(401), 464–77.

Arain, A. L., Pummill, R., Adu-Brimpong, J., Becker, S., Green, M., Ilardi, M. and Neitzel, R. L. (2020). Analysis of e-waste recycling behavior based on a survey at a Midwestern US University. *Waste Management*, *105*, 119–27.

Bhutto, M. Y., Liu, X., Soomro, Y. A., Ertz, M., and Baeshen, Y. (2021). Adoption of energy-efficient home appliances: Extending the theory of planned behavior. *Sustainability*, *13*(1), 250.

Botetzagias, I., Dima, A. F. and Malesios, C. (2015). Extending the theory of planned behavior in the context of recycling: The role of moral norms and demographic predictors. *Resources, Conservation and Recycling*, *95*, 58–67.

Cai, S., Long, X., Li, L., Liang, H., Wang, Q., and Ding, X. (2019). Determinants of intention and behavior of low carbon commuting through bicycle-sharing in China. *Journal of cleaner production*, *212*, 602–09.

Chen, C. F., Xu, X. and Frey, S. (2016). Who wants solar water heaters and alternative fuel vehicles? Assessing social–psychological predictors of adoption intention and policy support in China. *Energy Research & Social Science, 15*, 1–11.

Chen, M. F. and Tung, P. J. (2014). Developing an extended theory of planned behavior model to predict consumers' intention to visit green hotels. *International Journal of Hospitality Management, 36*, 221–30.

Cho, Y. N., Thyroff, A., Rapert, M. I., Park, S. Y., and Lee, H. J. (2013). To be or not to be green: Exploring individualism and collectivism as antecedents of environmental behavior. *Journal of Business Research, 66*(8), 1052–59.

De Leeuw, A., Valois, P., Ajzen, I. and Schmidt, P. (2015). Using the theory of planned behavior to identify key beliefs underlying pro-environmental behavior in high-school students: Implications for educational interventions. *Journal of Environmental Psychology, 42*, 128–38.

Dhir, A., Koshta, N., Goyal, R. K., Sakashita, M., and Almotairi, M. (2021). Behavioral reasoning theory (BRT) perspectives on E-waste recycling and management. *Journal of Cleaner Production, 280*, 124–69.

ElHaffar, G., Durif, F., and Dubé, L. (2020). Towards closing the attitude-intention-behavior gap in green consumption: A narrative review of the literature and an overview of future research directions. *Journal of cleaner production, 275*, 122–56.

Fornell, C. and Larcker, D. F. (1981). Evaluating structural equation models with unobservable variables and measurement error, *Journal of Marketing Research, 18*(1): 39–50.

Forti, V., Balde, C. P., Kuehr, R., and Bel, G. (2020). The Global E-waste Monitor 2020: Quantities, flows, and the circular economy potential.

Gao, L., Wang, S., Li, J. and Li, H. (2017). Application of the extended theory of planned behavior to understand an individual's energy-saving behavior in workplaces. *Resources, Conservation and Recycling, 127*, 107–13.

Hair Jr, J. F., Hult, G. T. M., Ringle, C., and Sarstedt, M. (2016). A primer on partial least squares structural equation modeling (PLS-SEM) (2nd ed.). Thousand Oaks, United States Sage Publications.

Hair Jr, J. F., Sarstedt, M., Ringle, C., and Mena J.A. (2012). An assessment of the use of partial least squares structural equation modeling in marketing research. *Journal of the academic marketing science*. 40/3, 414–33.

Han, H., and Yoon, H. J. (2015). Hotel customers' environmentally responsible behavioral intention: Impact of key constructs on the decision in green consumerism. *International Journal of Hospitality Management, 45*, 22–33.

Hartmann, P., Eisend, M., Apaolaza, V., and D'Souza, C. (2017). Warm glow vs. altruistic values: How important is an intrinsic emotional reward in pro-environmental behavior? *Journal of Environmental Psychology, 52*, 43–55.

Hu, L. T., and Bentler, P. M. (1999). Cutoff criteria for fit indexes in covariance structure analysis: Conventional criteria versus new alternatives. *Structural Equation Modeling: A Multidisciplinary Journal, 6*(1), 1–55.

Hulland, J. (1999). Use of partial least squares (PLS) in strategic management research: A review of four recent studies. *Strategic Management Journal, 20*(2), 195–204.

Isen, A. M. (1970). Success, failure, attention, and reaction to others: The warm glow of success. *Journal of personality and social psychology, 15*(4), 294.

Judge, M., Warren-Myers, G., and Paladino, A. (2019). Using the theory of planned behavior to predict intentions to purchase sustainable housing. *Journal of cleaner production, 215*, 259–67.

Kinnear, T. C., Taylor, J. R., and Ahmed, S. A. (1974). Ecologically concerned consumers: who are they? Ecologically concerned consumers can be identified. *Journal of Marketing, 38*(2), 20–24.

Kumar, A. (2019). Exploring young adults' e-waste recycling behavior using an extended theory of planned behavior model: A cross-cultural study. *Resources, Conservation and Recycling, 141*, 378–89.

Lee, Y. K., Kim, S., Kim, M. S., and Choi, J. G. (2014). Antecedents and interrelationships of three types of pro-environmental behavior. *Journal of Business Research, 67*(10), 2097–2105.

Lowe, B., Lynch, D. and Lowe, J. (2015). Reducing household water consumption: a social marketing approach. *Journal of Marketing Management, 31*(3-4), 378–408.

Nguyen T. N., Nguyen, H. V., Lobo, A. and Dao T. S. (2017). Encouraging Vietnamese household recycling behavior: Insights and implications. *Sustainability, 9*(2), 179.

Nguyen, T. N., Lobo, A., and Greenland, S. (2016). Pro-environmental purchase behavior: The role of consumers.

Nie, H., Vasseur, V., Fan, Y., and Xu, J. (2019). Exploring reasons behind careful use, energy-saving behaviors in the residential sector based on the theory of planned behavior: Evidence from Changchun, China. *Journal of Cleaner Production, 230*, 29–37.

Nimri, R., Patiar, A., and Jin, X. (2020). The determinants of consumers' intention of purchasing green hotel accommodation: Extending the theory of planned behavior. *Journal of Hospitality and Tourism Management, 45*, 535–43.

Perugini, M. and Bagozzi, R. P. (2001). The role of desires and anticipated emotions in goal-directed behaviors: Broadening and deepening the theory of planned behavior. *British Journal of Social Psychology, 40*(1), 79–98.

Scalco, A., Noventa, S., Sartori, R., and Ceschi, A. (2017). Predicting organic food consumption: A meta-analytic structural equation model based on the theory of planned behavior. *Appetite, 112*, 235–48.

Si, H., Shi, J. G., Tang, D., Wu, G., and Lan, J. (2020). Understanding intention and behavior toward sustainable usage of bike-sharing by extending the theory of planned behavior. *Resources, Conservation and Recycling, 152*, 104–513.

Tan, C. S., Ooi, H. Y. and Goh, Y. N. (2017). A moral extension of the theory of planned behavior to predict consumers' purchase intention for energy-efficient household appliances in Malaysia. *Energy Policy*, 107, 459–71.

Taufik, D., Bolderdijk, J. W., and Steg, L. (2015). Acting green elicits a literal warm glow. *Nature Climate Change*, *5*(1), 37–40.

Truong Y. and McColl, R. (2011). Intrinsic motivations, self esteem, and luxury goods consumption. *Journal of Retailing and Consumer Services*, *18*(6), 555–61.

Verma, V. K and Chandra, B. (2018). An application of the theory of planned behavior to predict young Indian consumers' green hotel visit intention. *Journal of Cleaner Production*, *172*, 1152–62.

Wang, B., Ren, C., Dong, X., Zhang, B., and Wang, Z. (2019). Determinants shaping willingness towards on-line recycling behavior: An empirical study of household e-waste recycling in China. *Resources, Conservation and Recycling*, *143*, 218–25.

Wang, Z., Guo, D., and Wang, X. (2016). Determinants of residents' e-waste recycling behavior intentions: evidence from China. *Journal of cleaner production*, *137*, 850–60.

Wang, Z., Guo, D., Wang, X., Zhang, B., and Wang, B. (2018). How does information publicity influence residents' behavior intentions around e-waste recycling? *Resources, Conservation and Recycling*, *133*, 1–9.

Yadav, R., and Pathak, G. S. (2016). Young consumers' intention towards buying green products in a developing nation: Extending the theory of planned behavior. *Journal of Cleaner Production*, *135*, 732–39.

Yu, S., and Lee, J. (2019). The effects of consumers' perceived values on intention to purchase upcycled products. *Sustainability*, *11*(4), 1034.

Zhang, B., Du, Z., Wang, B., and Wang, Z. (2019). Motivation and challenges for e-commerce in e-waste recycling under "Big data" context: a perspective from household willingness in China. *Technological Forecasting and Social Change*, *144*, 436–44.

An Assessment on Consumer Dynamics for Green Homes 6

OZAN ATIK AND NILGÜN GÜRKAYNAK

Introduction

The energy crisis of the 1970s led to an awareness on environmental health, efficient resource usage and clean energy. In the following decades, the term "green consumption" was coined to define consumption patterns that safeguards the environment for the present and the future (Zhang, Kang, and Jin, 2018).

The increasing population and industrialization rise the demand for energy, mainly from fossil fuels, leading to higher carbon emissions, which in turn cause global warming and climate change. Seventy-nine percent of the primary energy consumption in the world is met by fossil fuels. In Turkey, this rate is around 83 percent. According to the 2019 Global Status Report for Buildings and Construction, buildings and construction are responsible for 38 percent of all carbon emissions in the World (Global Alliance for Buildings and Construction, 2020). With global warming and climate change reaching alarming proportions in recent years, the Paris Agreement was signed in 2015. With the Paris Agreement, it is aimed to reduce worldwide carbon emissions by 50 percent by 2030 and to zero by 2050.

Considering the 38 percent share of carbon emission by buildings and the respective effect on the environment, the way buildings—especially the residential ones are designed—requires special attention. Green buildings can decrease, or even, eliminate such unfavorable impacts. By understanding the consumer awareness, attitudes, and readiness for green buildings constructions firms, professional associations, public organizations will be able to strategize better. Understanding green consumers and their characteristics are getting crucial for all these stakeholders. While some consumers are willing to purchase a green home, others are not, which suggests that there are individual differences among consumers about factors affecting decision making (Haws, Winterich, and Naylor, 2014).

The primary aim of this study is to identify the factors that facilitate the purchase green homes. The study investigates the effects of EC, EA, and SN of people who plan to purchase a home in the near future.

This study is based on a model in which the mediating effect of attitude and the moderator effect of PS on PI are discussed together. Earlier studies discussed moderating effect of PS or mediating effect of the attitude. Almost all studies on green building in Turkey have been handled in different academic fields, especially in architecture. This is one of the pioneering studies in the field of marketing in Turkey.

Following a brief review of previous studies and conceptual definitions, hypotheses of the study, adopted methodology, findings, and their implications for various stakeholders in construction sector will be presented.

Literature Review and Hypothesis Development

Green Home

Result of the oil shocks of the early 1970s, marked the first serious concern about resources. The result was a solar technology, explosion of interest in energy efficiency, retrofitting homes and commercial buildings with insulation, and energy recovery systems (Kibert, 2004). Subsequently, the British green building system, the Building Research Establishment Environmental Assessment Method (BREEAM), was developed in 1992. In 1993, inside the boardroom of the AIA (American Institute of Architect), with representatives from over sixty firms and nonprofits the U.S. Green Building Council (USGBC) was founded. Then, USGBC had developed Leadership in Energy and Environmental Design (LEED) to measure the sustainability and performance of a building.

Green buildings are considered as a solution to overcome environmental problems such as extreme energy consumption and air pollution (Retzlaff, 2009). Main goals of the green building/home are enhancing energy efficiency, utilizing durable and non-toxic materials, and reducing water consumption with high-recycled content materials (Ali and Nsairat, 2009). Kubba (2010) states that green building about enhances the efficiency in building by using energy, water, and materials to protect natural resources and improve built environment.

Purchase Intention (PI)

Intention stems from psychology and is considered a type of personal motivation that can accurately estimate individual behavior (Eagly and Chaiken, 1993). Intention refers to whether an individual is ready to perform a particular behavior (Yadav and Pathak, 2016). Ajzen and Fishbein (1975) became the first to refer to consumers' intention to buy certain products or services. Ajzen's Theory of Planned Behavior (TPB) basically assumes that an individual's intention to perform a behavior is influenced by three parameters: attitude toward the behavior, SN, and perceived behavioral control over the behavior's performance (Ajzen, 1991). Many different definitions of purchase intention (PI) have been made in the literature so far. According to Rosa and others (2006) PI is an important indicator in predicting consumers' future behavior. Teng and others (2013) defined PI as the prospective of a consumer who is planning to buy a product in the future. Shah and others (2012) defined that, PI refers to the purchasing decision-making process in which the consumer takes into account various conditions and reasons. Evaluating an individual's PI is a complex process, which is directly related to the consumer's perceptions, behaviors and attitudes toward the product or service itself or even toward the seller (Huete Alcocer, 2017).

People are more environmentally conscious than before. Many of them have concerns about the environment and therefore consumers intend to buy environmentally friendly products that are less destructive (Peattie, 1995). With the increasing green consumption demands, it has become very important to understand the green PI of the consumer. Green PI has been discussed by many researchers in the literature. Green PI is conceptualized as the possibility and desire of a person to prefer environmentally friendly products over other traditional products in their purchasing decision (Abdul Rashid, 2009). Chan and others (2001) defined green purchase as a specific type of environmentally friendly behavior that consumers display to express their concerns about the environment. Ramayah and others (2010) identified the green PI as representative of the real purchase.

Environmental Concern (EC)

EC is an emotional trait that expresses an individual's worries, thoughts, likes and dislikes about what affects the environment (Yeung, 2004). EC has been evaluated as predictive in understanding environmental behavior (Ajzen and Fishbein, 1977). Schultz and others (2000) determined that EC has three interrelated factors: ego (egoistic), biosphere (biosphere), and other people (altruistic). SGuin and others (1998) have stated that EC can have a significant effect on motivation level of the individuals' attempts toward environmental problems. According to Zimmer and others (1994), dimensions of the EC consist of sensitivity to climate change issues, energy conservation, awareness of alternative energy sources and clean energy. Stern (1992) established a view that EC is a function of a deeper cause such as basic religious beliefs or post-materialistic values. Dunlap and Van Liere (1978) developed the New Ecological Paradigm (NEP) scale, which is admitted as the oldest quantitative definition of EC. People with high levels of EC tend to respond to environmental problems and can act faster to protect the environment (McDonald et al., 2015). Previous studies stated that green PI is positively affected by EC directly (Han, Hsu, and Sheu, 2010; Aman, Harun, and Hussein, 2012; Tan, 2013; Suki, 2013; Sang and Bekhet, 2015; Zhang, Kang, and Jin, 2018; Xu, Hua, Wang, and Xu, 2020). Other studies also emphasized that, EC also influence the attitudes towards green product (Mostafa, 2007; Hanson, 2013; Yadav and Pathak, 2016; Maichum, Parichatnon, and Peng, 2016). Hence, the following hypotheses are proposed:

H1: EC has a significant positive impact on consumers' PI.
H2: EC has a significant positive impact on consumers' attitude towards purchasing a green home.

Environmental Awareness (EA)

EA plays a significant role in pursuing green activism. EA has been defined by the researchers by means of different words that will have the same meaning. Generally, EA is defined as knowledge about the general concept and fact related to the ecosystems and environment (Mostafa M. M., 2007). This awareness is harmonious with the view that the world's natural resources are limited, and the ecological balance may be at a level of significant deterioration that requires urgent action (Hayes, 1990). EA is also defined as being aware of environmental problems and active participation in environmental organizations (Altin, Tecer, Tecer, Altin, and Kahraman, 2014). According to Karatekin (2014) EA is the trigger of positive attitudes towards environmental behavior. Carlson (2004) stated that EA is the first stage of becoming a green consumer. According to Carlson, even if people are environmentally aware, they don't necessarily have to behave in an environmentally friendly way. EA consists of a positive attitude and behavior toward the environment. According to Axelrod (1994), EA is the most crucial part of behavioral science because it includes environmental attitude, beliefs, and values. EA plays an important role in specifying the intention to purchase green product. Jaiswal and Kant (2018) stated that consumer EA is significant because it can explain consumers' attitudes towards environmentally friendly products. A study conducted in California revealed that a high level of EA leads to a high level of PI of hybrid electric vehicles (Heffner, Kurani, and Turrentine, 2007). A study among young Chinese consumers showed that awareness positively moderates association between attitude and PI for organic food (Ahmed et al., 2019). Kumar and others (2017) found a weak relationship between EA and PI. In another study, Aman and others (2012) found that although EA positively influences the PI, there is no link between EA and consumer attitude toward green products. In United Arab Emirates context, Khaleeli and Jawabri (2021) found a positive and significant relationship between EA and consumers' attitudes toward environmentally friendly products. Hence, the following hypotheses are proposed:

H3: EA has a significant positive impact on consumers' PI.
H4: EA has a significant positive impact on consumers' attitude towards purchasing a green home.

Subjective Norm (SN)

SN is the social pressure an individual takes while undertaking a specific behavior (Ajzen, 1991). Social pressure comes mostly from the attitudes or views of others or groups that are important to a person as to whether the person should perform a certain behavior (Park, 2000). Dean and others (2012) argued that positive perception of important others like family, relatives and friends have an impact on green PI of an individual. According to Zukin and Maguire (2004), SN has a major influence on green consumption. Sreen and others (2018) state that there is a path from SN to attitude toward green products and then to PI. While there are many studies on SN and PI in the literature, there are few studies on testing the relationship between SN and attitude toward purchasing a green product. In a study with young consumers in India, Varshneya and others (2017) found that there was no significant relationship between SN and the attitude toward using organic clothing. In contrast to this study, another study conducted in India revealed a direct significant relationship between SN and attitude toward green product (Sreen, Purbey, and Sadarangania, 2018). Previous studies have also stated that SN is the main determinant of PI (Sparks and Shepherd, 1992; Kaiser and Gutscher, 2003; Zukin and Maguire, 2004; Han, Hsu, and Sheu, 2010; Dean, Raats, and Stepherd, 2012; Moons and Pelsmacker, 2012; Teng, Wu, and Liu, 2013; Paul, Modi, and Patel,

2016; Yadav and Pathak, 2016; Chaudhary and Bisai, 2018). On the other hand, In Indian context, Khare (2015) revealed that social norms did not influence green buying behavior. Hence, the following hypotheses are proposed:

H5: SN has a significant positive impact on consumers' PI.
H6: SN has a significant positive impact on consumers' attitude towards purchasing a green home.

Attitude Toward Purchase of Green Home (ATTGH)

Attitude is admitted one of the most significant impacts on behavior. Spruyt and others (2007) stated that attitude is a significant factor in understanding behavior and is related to their knowledge and personal experiences. Kotchen and Reiling (2000) defined environmental attitudes as predictors of environmentally based action. According to Ajzen and Fishbein (1977), behavior and attitudes should be measured at the same level of compliance. There are several theories put forward to explain the process of attitudes predicting behavior. The most important one of these is the theory of reasoned action (TRA) developed by Martin Fishbein and Icek Ajzen in 1967. The theory is mainly used to predict how individuals will behave according to their pre-existing attitudes and behavioral intentions. Hence, according to the theory, people's attitudes play an important role when it comes to establishing the intention to engage in a certain behavior. Past studies have shown that consumer attitude toward green products significantly affects green PI (Aman, Harun, and Hussein, 2012; Yadav and Pathak, 2016; Mostafa M., 2009). Most of the previous studies used attitudes as a direct variable influencing PI along with EC, EA, and SN. Some studies also used attitude toward green product as a mediating variable. These studies revealed that there is a path from subjective norms to attitude toward green products and then to PI (Tarkiainen and Sundqvist, 2005; Sreen, Purbey, and Sadarangania, 2018). Paul and others (2016) determined that important direct and indirect effect of EC on green PI through the mediation of TPB variables. Li and others (2019) showed that EC influences PI indirectly with mediation role of attitude. A study among young Chinese consumers showed that awareness positively moderates association between attitude and PI for organic food (Ahmed et al., 2019). Aman and others (2012) found that although EA positively influences the PI, there is no link between EA and consumer attitude toward green products. Khaleeli and Jawabri (2021) found that attitudes did not have a mediating role in the relationship between awareness and PI. Hence, the following hypotheses are proposed:

H7: ATTGH has a significant positive impact on consumers' PI.
H8: ATTGH mediates the relationship between EC, EA, SN, and PI.
H8a: ATTGH mediates the relationship between EC and PI.
H8b: ATTGH mediates the relationship between EA and PI.
H8c: ATTGH mediates the relationship between SN and PI.

Price Sensitivity (PS)

Marketers are often interested in evaluating the price sensitivities of consumers to make strategic and tactical decisions about market segmentation, marketing activities, pricing, and competitive marketing strategies (Kaul and Wittink, 1995). According to Goldsmith and others (2005), price sensitivity (PS) relates to how consumers react to price levels and price changes. Anderson (1996) defined that PS is related to the extent to which the consumer accepts a price increase for a certain product due to economic and psychological gains. Price-sensitive consumers will be less willing to pay higher prices for the same goods than less price-sensitive consumers (Foxall and James, 2003). Price plays an important role in the purchasing decision of consumers. Perceptions about quality and PI may differ between consumers who are price-sensitive and insensitive. Price sensitive consumers may take price as the most important consideration in purchasing decisions. Thus, it is very important that marketers try to understand the PS of target customers and developing strategies (Goldsmith, Kim, Flynn, and Kim, 2005). Studies demonstrate that PS plays a significant role in increasing the intention to purchase toward green products. A study conducted by Ng and others (2018), which measures the PI of electric vehicles, showed that PS is positively associated with PI. Hahnel and others (2014) found that PS was an important factor influencing the consumer's PI, and that consumers with lower PS were more likely to pay more for electric vehicles than conventional vehicles. Most studies view PS as a direct or indirect predecessor to the PI of an environmentally friendly product. Some researchers explored the moderator role between consumers' EA and EC and green PI (Erdil, 2018). Hsu and others (2017) proved that PS plays a moderator role in the relationship between PI of green skincare products and attitude. Likewise, Bhutto and others (2020) used PS as a moderator in the study on the PI of hybrid vehicles. Hence, the following hypothesis is proposed:

H9: PS plays a negative moderating role between ATTGH and PI.

Figure 6.1 shows the proposed hypothesized research model.

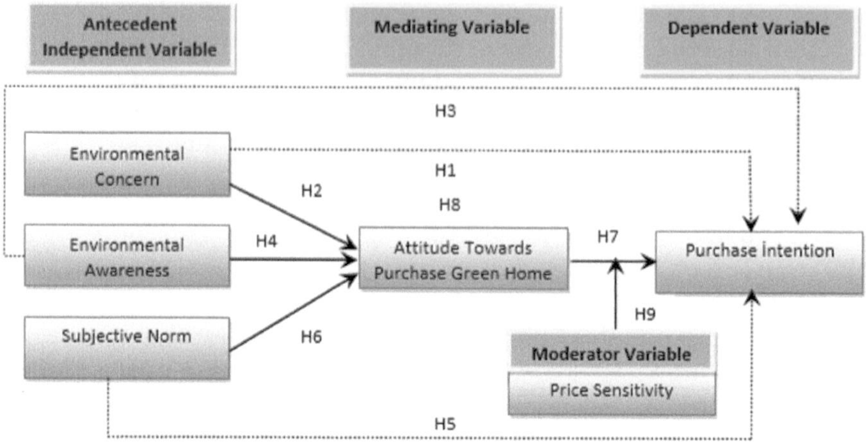

Figure 6.1. Proposed hypothesized research model
Source: Authors preparation

Methodology

Data Collection, Sampling

The universe of this study is potential home buyers in Turkey. In 2020, 1,499,316 houses were sold in the country. Of the houses sold, 469,740 were the first sales. According to these data, 17.68 percent of these sales took place in Istanbul, 10.48 percent in Ankara and 6.23 percent in Izmir. In Turkey, 518 green building certified projects have been completed so far. Three hundred and forty-four of these projects were carried out within the borders of Istanbul and forty-five of them in Izmir and Ankara provinces (https://cedbik.org/tr/sertifikali-projeler, 2021). Therefore, based on these data, those living in Istanbul, Ankara, and Izmir, which have a housing purchase plan, were selected for the sample that will represent the universe of the study.

The fieldwork was conducted online through Google Forms. A pilot study was conducted with fifteen persons before starting the fieldwork of the survey. As a result of the pilot study, wording of some questions has changed, and explanations have been added to some parts to reduce the risk of misunderstanding.

The initial sample for the survey included 332 people, ninety-two of which were excluded in the analysis as they either did not plan to buy a home in the short or long term or resided in other locations, leading to final sample of 240. SPSS 26 software was used for statistical analyses.

Questionnaire Design

Since the sample of the research is those who plan to buy a home in the future, the first question of the survey asked whether they were considering buying a home in the short or long term. After that, twelve items were prepared in which the characteristics of the green home were listed in order to measure the knowledge level of the participants about the green home. Respondents were asked which ones are one of the characteristics of a green home. Twelve items were created by taking expert opinions from the sector and green building associations. After measuring the knowledge level of consumers about green buildings, a brief description was given about green homes before starting the other survey questions, as some respondents may not be aware of the green home concept.

The items used to measure EC were adapted from Dunlap and Van Liere (1978). This scale, which has been used in environmental issues since 1978, has been named the New Environmental Paradigm (NEP). The scale, which consists of twelve Likert-type questions developed by Dunlap and Van Liere (1978), was later revised by Dunlap and others (2000) by adding new components to the scale. A Likert-type five-point rating was used in the scale consisting of fifteen items. Only four items were taken from this scale to measure EC. Nine items used to measure EA were adapted from Berberoğlu and Uygun (2012). The items of the other independent variable, subjective norms, were adapted from Chen and Peng (2012), Liobikienė and others (2016), and Dean and others (2011). The SN scale consists of two items. A one-item, five-point Likert-type scale was operationalized to measure ATTGH purchase based on Taylor and Todd (1995), Chan (2001), and Mostafa (2009).

The scale used by Goldsmith and others (2005) to measure the PS of new fashion clothes was used to measure the PS of green homes. In addition, two items were adapted from the Lichtenstein and others (1993) and Sinha and others (1999). The scale used to measure PS consists of five items. To measure the price perception of green homes, the following question was asked: "How much more expensive do you think a home designed in accordance with the green home concept is than other homes?" The next question was "In how many years do you expect the excess amount you will pay for a green home to amortize itself?" Two items used to measure PI toward green homes were adopted from Yee (1997), Mostafa (2009), and Taylor and Todd (1995). All items used to measure independent and dependent variables were designed using a Likert scale ranging from 1 (strongly disagree) to 5 (strongly agree). The table below

Table 6.1. Items Used in the Survey

Variables	*Items*	*Reference*
Environmental concern	4	(Dunlap, Liere, Mertig, & Jones, 2000)
Environmental awareness	9	(Berberoğlu & Uygun, 2012)
Subjective norm	2	(Chen & Peng, 2012; Liobikienė, Mandravickaitė, & Bernatonienė, 2016; Dean, Raats, & Shepherd, 2011)
Attitude towards green home	1	(Taylor & Todd, 1995; Chan, 2001; Mostafa M., 2009)
Price sensitivity	5	(Goldsmith, Kim, Flynn, & Kim, 2005)
Purchase intention	2	(Yee, 1997; Mostafa M., 2009; Taylor & Todd, 1995)

Source: Authors preparation

shows the sources and numbers of the questions used in the questionnaire of this research, according to the variables (Table 6.1). Finally, the questionnaire was completed with demographic questions.

Findings

Descriptive Analysis

Of the 240 respondents, 58.8 percent (141) were female, 41.3 percent (ninety-nine) were male; 56.3 percent were single, 44.1 percent were married. A majority of the respondents—66.7 percent (160)—were between twenty-five and forty-four years old; 31.3 percent (75) were with income level $501–800; 53.3 percent (128) were university graduates; 71.3 percent (111) had A and B social status (Table 6.2).

Table 6.2. The Descriptive Analysis of Respondents

Gender	*Frequency*	*Percent*	*Marital Status*	*Frequency*	*Percent*
Female	141	58,8	Married	135	56,3
Male	99	41,3	Single	105	43,8
Total	240	100,0	Total	240	100,0
Age	*Frequency*	*Percent*	*Social Status*	*Frequency*	*Percent*
18–24 years	27	11,3	D	5	2,1
25–34 years	75	31,3	C2	28	11,7
35–44 years	85	35,4	C1	36	15,0
45–54 years	40	16,7	B	87	36,3
55 years and more	13	5,4	A	84	35,0
Total	240	100,0	Total	240	100,0

Income Level	*Frequency*	*Percent*	*Education*	*Frequency*	*Percent*
0–200 $	28	11,7	Primary School	2	0,8
201–500 $	53	22,1	Secondary School	3	1,3
501–800 $	75	31,3	High School	24	10,0
801–1100 $	50	20,8	Associate Degree (2 years)	24	10,0
1101–1400 $	14	5,8	University (distance education)	18	7,5
1401 $ and more	20	8,3	University	128	53,3
			Postgraduate	41	17,1
Total	240	100,0	Total	240	100,0

Source: Authors preparation

Seventy percent (168) of the participants have no knowledge of the green home. While 22.5 percent (54) of the participants have some knowledge, only 7.5 percent (18) claim to know the green home concept (see Table 6.3).

Table 6.3. Do you know the green home concept?

	Frequency	*Percent*	*Cumulative Percent*
I have never heard it before	124	51,7	51,7
I heard, I don't know	44	18,3	70,0
I have some knowledge	54	22,5	92,5
Yes, I know the green home concept	18	7,5	100,0
Total	240	100,0	

Source: Authors calculations

The participants were asked which of the following items is one of the characteristics of a house designed in accordance with the green home concept. In addition to avoiding the factors that cause environmental pollution, energy-saving and effective usage of daylight are the most well-known features of the green home. Proximity of the building location to public transport and to social life and good lighting are the least known features of it (see Table 6.4).

Table 6.4. Which of the following items do you think is one of the characteristics of a green home?

	Responses		
	N	*Percent*	*Percent of Cases*
The proximity of the building location to social life	59	3,8	24,6
The proximity of the building location to public transport	22	1,4	9,2
Energy-saving	188	12,2	78,3
Water efficiency	144	9,3	60,0
Low carbon footprint	139	9,0	57,9
Indoor air quality	114	7,4	47,5
Good lighting	38	2,5	15,8
Effective usage of daylight	171	11,1	71,3
Avoiding the factors that cause environmental pollution during the construction phase of the building	191	12,4	79,6
The recyclability of the materials used in the building construction process.	151	9,8	62,9
Energy efficiency	159	10,3	66,3
Usage of renewable energy sources	168	10,9	70,0
	1544	100,0	643,3

Source: Authors calculations

In another question, respondents were asked how much more expensive a green home was than other homes. According to the answers given, 54.6 percent of the respondents think it is 10 percent or more expensive. The rate of those who consider that it is more expensive between 1 and 3 percent was only 4.2 percent (see Table 6.5).

After this question, participants who thought it was more expensive were asked "In how many years do you expect the excess amount you will pay for a green home to amortize itself?" The participants in the "No price difference" section consist of participants who consider that there is no dif-

Table 6.5. How much more expensive do you think a home designed in accordance with the green home concept is than other homes?

	Frequency	*Percent*	*Cumulative Percent*
No difference	10	4.2	4.2
1–3% more expensive	22	9.2	13.3
4–6% more expensive	36	15.0	28.3
7–9% more expensive	41	17.1	45.4
10% or more	131	54.6	100.0
Total	240	100.0	

Source: Authors calculations

Table 6.6. In how many years do you expect the excess amount you will pay for a green home to amortize itself?

	Frequency	*Percent*	*Cumulative Percent*
No price difference	10	4.2	4.2
In 1–2 years	33	13.8	17.9
In 3–4 years	65	27.1	45.0
In 5–6 years	68	28.3	73.3
In 7–8 years	31	12.9	86.3
9 years and more	33	13.8	100.0
Total	240	100.0	

Source: Authors calculations

ference in price in green home compared to other homes in the previous question. In this question, while 28.3 percent of the participants expect it to amortize for itself in 5 to 6 years, 27.1 percent expect it to amortize for itself in 3 to 4 years (Table 6.6).

Data Analysis

Primarily, factor and reliability analysis of the scales should be performed, before testing of the hypotheses. Before the start exploratory factor analysis, Kaiser Meyer Olkin (KMO) and Bartlett tests, which show the suitability of the sample data for factor analysis, were performed and appropriate results were produced (KMO = 0.916, Bartlett = 0.000). The fact that the KMO value takes a value between 0.5 and 1 indicates that there is a necessary sample size for factor analysis. For the KMO value, below 0.5 is "unacceptable," between 0.5 and 0.7 is "mediocre," between 0.7 and 0.8 is "good," between 0.8 and 0.9 is "great," and 0.9 is rated as "superb" for above (Field, 2007). Bartlett's Test of Sphericity result being at $p<0.05$ level indicates that the relationship between the variables is significant (Kalaycı, 2010). Thus, exploratory factor analysis yielded 5 different factors.

EC2 (environmental concern) and PS1 (Price Sensitivity) were excluded from the factor analysis due to double loading. The EC4 (If things continue their present course, we will soon experience a major ecological catastrophe) item, which is included in the EC scale, was loaded on the environmental awareness factor at a high rate of 0.900. These items, which were used to measure EC in studies conducted in other countries, were confirmed in the environmental awareness scale in this study conducted in Turkey.

According to Hair, Babin, Money, and Samouel (2003), Cronbach's alpha value greater than 0.60 is acceptable and considered good. In addition, to measure convergent validity, Composite Reliability (CR) and Average Variance Extracted (AVE) tests were performed (Table 6.7). The CR value was between 0.733 and 0.970, and the AVE value was between 0.582 and 0.760. In the literature, it is accepted that the AVE value should be greater than 0.5 and the CR value should be greater than 0.7. Thus, these results revealed the reliability and internal consistency of the scales (Hair, Black, Babin, and Anderson, 2010).

Initially, multiple linear regression analysis was done to attitude toward purchase green home as the dependent variable, and EA, EC, and SN as the independent variables. Before multiple regression analysis, it is necessary to test whether there is a multicollinearity problem. Multicollinearity occurs when an independent variable is highly correlated with one or more of the other independent variables in a multiple regression equation (Allen, 1997). In general, it is considered appropriate that the correlation between independent variables is below 0.80 (Berry and Feldmen, 1985). Another

Table 6.7. Factor Loadings, Reliability Analysis, AVE and CR

Construct	*Items*	*Factor Loadings*	*Factor*	*Cronbach's Alpha*	*AVE*	*CR*
Environmental Awareness	EA4	0,963	50.38%	0,973	0,760	0,970
	EA1	0,929				
	EA2	0,926				
	EA3	0,906				
	EC4	0,9				
	EA5	0,898				
	EA9	0,864				
	EA7	0,805				
	EA8	0,778				
	EA6	0,769				
Price Sensitivity	PS3	0,86	10.53%	0,815	0,611	0,860
	PS2	0,852				
	PS5	0,785				
	PS4	0,603				
Purchase Intention	PI2	0,976	9.30%	0,928	0,941	0,970
	PI1	0,965				
Environmental Concern	EC3	0,851	6.18%	0,623	0,582	0,733
	EC1	0,664				
Subjective Norm	SN1	0,923	3.87%	0,822	0,717	0,834
	SN2	0,764				

Source: Authors calculations

Table 6.8. Regression Analysis

Hypothesis	*Path*	*Beta*	*t*	*Result*
H1	EC–PI	–0,051	–0,741	Not Supported
H2	SN–PI	0,392	5,551*	Supported
H3	EA–PI	0,139	1,945	Not Supported
H4	EC–ATTGH	–0,024	–0,453	Not Supported
H5	SN–ATTGH	0,557	10,253*	Supported
H6	EA–ATTGH	0,280	5,095*	Supported
H7	ATTGH–PI	0,485	7,455*	Supported

Note. EA = Environmental Awareness, PS = Price Sensitivity, PI = Purchase Intention, EC = Environmental Concern, SN = Subjective Norm, ATTGH = Attitude Towards Purchase Green Home, * $p < 0{,}05$.

Source: Authors calculations

way to test the multicollinearity problem is to check the VIF value. There are sources in the literature showing that this value should be below 2.5 (Allison, 1999), 5 (Craney and Surles, 2002), and 10 (Hair, Anderson, Tatham, and Black, 1995). Values were found to be appropriate for both analyzes (The Pearson Correlation between variables in this study were between 0.19 and 0.47).

Table 6.8 shows the direction and significance of the relationship of the independent variables on the dependent variable. Accordingly, a significant positive correlation was found between the SN and the PI (ß = 0.392, t = 5.55, $p < 0.001$). However, significant relationship was not found between the EC and the PI (ß = –0.051, t = –0.74, $p > 0.045$). Likewise, no significant relationship was found between EA and PI (ß = 0.139, t = 1.94, $p > 0.05$). Thus, while H2 was accepted, H1 and H3 were not.

Secondly, multiple linear regression analysis was done to PI as the dependent variable, and EA, EC and SN as the independent variables. The correlation between the independent variables is less than 0.80 and the VIF value is less than three. Therefore, there is no multicollinearity problem in the model. Table 6.9 also shows the direction and significance of the relationship of the independent variables on the dependent variable. Accordingly, a significant positive correlation was found between the SN and the ATTGH (ß = 0.557, t = 10.25, $p < 0.001$). Significant positive correlation was found between the EA and the ATTGH (ß = 0.280, t = 5.09, $p < 0.001$). However, no significant relationship was found between EC and ATTGH (ß = –0.024, t = –0.45, $p > 0.65$). Thus, while H5 and H6 were accepted, H4 was not.

Thirdly, Simple linear regression analysis was done to PI as the dependent variable, and ATTGH as the independent variables. a significant posi-

Table 6.9. Mediating Variable Analysis of Attitude Toward Green Home

Constant	*Coeff*		*p*	*LLCI*	*ULCI*
SN	.4196		.000	.1661	.0490
SN	.2535		.0009	.1051	.4019
ATTGH	.2557		.0016	.0983	.4132
Mediator Variable	*Total Effect*	*Direct Effect*	*Indirect Effect*	*BootLLCI*	*BootULCI*
ATTGH	.4196	.2535	.1661	.0490	.3623

Note. SN = Subjective Norm ATTGH = Attitude Toward Purchase of Green Home

Source: Authors calculations

tive correlation was found between the ATTGH and the PI (ß = 0.435, t = 7.45, $p < 0.001$). Thus, H7 was accepted.

After regression analysis, the mediating role of ATTGH among independent variables was questioned using model 4 SPSS Process 3.5 (Table 6.9). In order for a variable to be considered a mediating variable, the independent variable must have a significant relationship with the dependent variable (Baron and Kenny, 1986). For this reason, a mediation test was not performed because there was no significant relationship between EC and PI, EA and PI in previous regression analyzes. Thus, hypotheses H7a and H7b were rejected without the need for testing.

The mediating variable may explain all or only a part of the observed relationship between the dependent and independent variables. The situation that it reflects the whole relationship is defined as complete mediation, and the situation that it reflects a part of it is defined as partial mediation. In the case of complete mediation, when the mediating variable is added to the analysis, the relationship between the dependent and independent variables is expected to weaken considerably or become statistically insignificant. In the case of partial mediation, the mediating variable cannot explain the entire relationship between the dependent and independent variables. The relationship between the dependent and independent variables continues to be significant, but there is a decrease in the level of significance (Yılmaz and Dalbudak, 2018). In addition, according to Preacher and Hayes (2004), the lower and upper confidence intervals obtained in the resampling (Bootstrap) method should not contain zero. As a result, both the lower and upper confidence intervals must be either positive or negative.

In Table 6.9, the mediating role of ATTGH between SN and PI is tested. First, it shows the influence of SN on PI alone. Secondly, the SN is shown together with the attitude variable. In this model, although the

Table 6.10. Moderating Variable Analysis of Price Sensitivity

Constant	*Coeff*	*p*	*LLCI*	*ULCI*
ATTGH	.5032	.0019	.1881	.8163
PS	–0,006	.9761	–0,3999	.3879
Int_1	–0,0141	.7680	–0,108	.0798

Note. PS = Price Sensitivity, ATTGH = Attitude Towards Purchase of Green Home

Source: Authors calculations

strength of the relationship between SN and intention to buy decreased, the level of significance remained. This shows that ATTGH has a partial mediating role between SN and PI. It was concluded that the mediating effect was significant because the bootstrap confidence intervals did not contain zero. Hence, H7c was accepted.

Table 6.10 shows the moderating role of PS among ATTGH and PI were questioned using model 1 SPSS Process 3.5. Although the model created to measure the regulatory effect was significant, PS and the interaction terms were not significant. Thus, hypothesis H9 that PS has a moderating effect is rejected.

Results, Discussion, and Implications

In this study, it was tested whether there is a relationship between consumers' EC, EA, SN and ATTGH, and PI, and whether attitude is the mediating variable and PS is the negative moderating variable.

First, the relationship between independent variables and PI toward green homes was tested. As a result of this regression analysis, only a significant relationship was found between SN and PI. It was not determined that EC and awareness had a significant effect on PI. As in this study, a significant relationship between SN and PI has been found in many studies in the literature (Sparks and Shepherd, 1992; Yadav and Pathak, 2016; Dean, Raats, and Shepherd, 2011; Han, Hsu, and Sheu, 2010). In addition, many studies have found a significant relationship between EA and PI (Heffner, Kurani, and Turrentine, 2007; Jaiswal and Kant, 2018). Unlike Kumar, K. Manrai, and Manrai (2017), who found a weak relationship between EA and PI. Although almost all studies found a significant relationship between EC and PI, surprisingly, no significant relationship was found in this study. As a result, in this study conducted in Turkey, it is understood that the PI is related to the behavior of people adapting themselves to social norms.

Second, the relationship between independent variables and ATTGH was tested. As a result of this regression analysis, a significant relationship was found between EA and ATTGH. Similarly Khaleeli and Jawabri (2021) found a significant relationship between SN and ATTGH, while Aman, Harun, and Hussein (2012) found no significant relationship. Although almost all studies found a significant relationship between EC and ATTGH, surprisingly, no significant relationship was found in this study.

Third, the mediating role of ATTGH between the SN and PI was tested. Since EC and EA are not directly and significantly related to PI, mediator variable analysis was not performed. As a result of the analysis, it was determined that ATTGH has a partial mediating role between SN and PI. Although the SN influences PI, relationship is strengthened via ATTGH.

Finally, the moderating effect of PS between ATTGH and PI was tested. Contrary to Hsu, Chang, and Yansritakul (2017), no significant relationship was found in this test. It was thought that PS would negatively affect PI. When the respondents were asked how expensive a green home was compared to other home, it was determined that 54.6 percent of the respondents had an expectation of more than 10 percent. It is considered as a positive result that the expected price difference of green homes is higher than the real price difference and that there is no decrease in the PI of price sensitive people.

According to the results, SN and EA have an important role in developing a positive ATTGH. The positive attitude also significantly increases the PI of the consumer. Contrary to many different countries, the fact that there is no negative effect on the PI of people with high PS, will support the studies on the concept of green home. For this reason, marketers, nongovernmental organizations and the government have a great responsibility. First of all, it is important to create EA in order to increase the intention to buy a green home. Business and marketing managers need to use engaging, educational marketing messages and slogans that combine the characteristics of these green home with desired consumer values. Avoiding environmental pollution during the construction of the building, energy saving and efficient use of daylight, which are among the most well-known features of the green home, should be emphasized more. In addition, marketers should be more educational and informative about one of the least known features of the green home, the proximity of the building location to public transportation and social life, about having good lighting. According to this study data, 51.7 percent of consumers have never heard of the green home concept before. If we take into account the people with have limited information knowledge, 92.5 percent of the consumers do not have enough information about the green home concept. For this reason, it is very significant for marketers, nongovernmental organizations and the government to implement educational and instructive marketing strategies.

The importance of the SN, which is the belief about the extent to which personal behaviors will be approved by others, and to what extent they conform to social behavior patterns, should not be denied. For this reason, the involvement of people valued by the society in marketing strategies will have a positive effect on raising awareness toward green home concept and raising PI and ATTGH.

Studies show that green buildings are between 1 and 2 percent more expensive than other homes (Kats, 2003). In addition, a study has calculated that the overpaid amount for the green building will pay for itself in six to seven years (World Green Building Trends 2018, 2018). According to the data of this study, 54.6 percent of consumers consider that green homes are 10 percent or above more expensive than other homes. When we take into account that the 7 to 10 percent excess amount that the consumers will pay for the green home does not have a negative effect on the PI, it is clearly seen that there is a situation that will create added value for the construction companies. With the spread of the green home concept, it is clear that consumers will be able to own these homes with a difference far below the price difference expectation range.

Implications for Construction Companies

Due to the high price expectations of consumers, there is a situation that will create added value for the construction companies. The approach of green home prices to conventional homes prices will increase the level of consumers' satisfaction and demand. Currently, almost all of the green buildings built in Turkey have the status of commercial buildings. For this reason, there is a serious potential for green residential buildings investments. In this context, the things to be done can be listed as follows:

- Marketing strategies should focus on "economies of green homes," well-known features of such as "energy saving" should be emphasized and consumer awareness should be created on the least known features such as the "good lighting." This will enable the construction companies to position their projects on what is valuable for the consumers.

- The involvement of people valued by the society in marketing strategies will have a positive effect on raising awareness toward green home concept and raising PI and ATTGH.
- It should be well explained to the consumers that the excess amount they will pay at the beginning will pay for itself in a short time.

Implications for Construction Materials Suppliers

The increasing number of green building projects creates opportunities for construction material suppliers. Recyclable and environmentally friendly products can be a priority on their product mix. The ecological and economic benefits of these products should be communicated to both constructors and potential purchasers of green homes through engaging, educational messages, and slogans. Carefully selected digital and social media marketing platforms need to be integrated with conventional mass media tools such as television and print. In addition, marketing managers can enhance their brand value by using PR tools such as sponsorship for scientific conferences, symposiums and participating in professional conventions and fairs.

Implications for NGOs and Professional Associations

NGOs and professional associations have an important role in informing and educating the public about a certain issue. These bodies have a higher message source credibility than commercial companies. As this study shows, we still have a long way to go to enlighten consumers about green buildings. For this reason, it is important for NGOs and professional associations to cultivate awareness and knowledge about green homes. These initiatives would aid the purchase intent toward ecologically friendly construction projects and home improvements. To accomplish these, we recommend the following:

- Many features of green buildings are unknown, online educational material need to be created by such bodies about the features and benefits of green homes.
- Free-access online calculation tool can be prepared to assess the amortization periods on green homes or eco-friendly home improvements.

Implications for Governments

Firstly, the most important factor that will expedite all these processes is the increase of the state's audit over the buildings and the preparation and implementation of the necessary legislation. Secondly, Broadcasting about creating environmental awareness in public announcement will form the infrastructure of creating a positive ATTGH.

Implications for Finance Sectors

The development of green finance is extremely important to ensure the sustainability of all these investments and to provide financing for green home/building projects.

Limitations and Future Research

The data of this study were collected only from Istanbul, Ankara, and Izmir provinces in Turkey. In this study, data collection was carried out using the nonprobability sampling technique. For this reason, education and social status levels were above the Turkey data averages. The majority of the participants in the study are university or postgraduate graduates and people with socioeconomic status levels A and B. The second limitation of the study is that there are many factors in the literature that affect the attitude and PI toward green homes, but in this study, only the effects of concern, awareness, SN factors on PI and ATTGH were examined. In future studies, many factors that are accepted in the literature that affect ATTGH and PI can be added to the research model. In addition, responses from rural areas regarding green homes can be surprising. Spreading the sample across Turkey may yield different results.

References

Abdul Rashid, N. N. (2009). Awareness of Eco-label in Malaysia's Green Marketing Initiative. In *International Journal of Business and Management* (pp. 132–41). CCSE.

Ahmed, N., Li, C., Khan, A., Qalati, S., Naz, S., and Rana, F. (2019). Purchase intention toward organic food among young consumers using theory of planned behavior: role of environmental concerns and environmental awareness. In *Journal of Environmental Planning and Management* (pp. 796–822). Routledge.

Ajzen, I., and Fishbein, M. (1977). Attitude-behavior relations: A theoretical analysis and review of empirical research. In *Psychological Bulletin* (pp. 888–918).

Ajzen, I. (1991). The Theory of Planned Behavior. *Organizational Behavior and HumanDecision Proceses*, 179–211.

Ajzen, I., and Fishbein, M. (1975). *Belief, attitude, intention, and*. Addison-Wesley.

Ali, H., and Nsairat, S. (2009). Developing a green building assessment tool for developing countries—Case. *Building and Environment*, 1053–64.

Allen, M. P. (1997). *Understanding Regression Analysis.* New York: Plenum Press.

Allison, P. D. (1999). *Multiple Regression: A primer.* London: Pine Forge Press.

Altin, A., Tecer, S., Tecer, L., Altin, S., and Kahraman, B. F. (2014). Environmental awareness level of secondary school students: A case study in Balıkesir (Türkiye). In *Procedia–Social and Behavioral Sciences* (pp. 1208–14). Elsevier.

Aman, A., Harun, A., and Hussein, Z. (2012). The Influence of Environmental Knowledge and Concern on Green Purchase Intention the Role of Attitude as a Mediating Variable. In *British Journal of Arts and Social Sciences* (pp. 145–67). British Journal Publishing.

Anderson, E. (1996). Customer Satisfaction and Price Tolerance. In *Marketing Letters* (pp. 265–74). Kluwer Academic Publishers.

Axelrod, L. (1994). Balancing Personal Needs with Environmental Preservation: Identifying the Values that Guide Decisions in Ecological Dilemmas. In *Journal of Social Issues* (pp. 85–104). Wiley.

Baron, R. M., and Kenny, D. A. (1986). The Moderator-Mediator Variable Distinction in Social Psychological Research: Conceptual, Strategic, and Statistical Considerations. In *Journal of Personality and Social Psychology vol.51* (pp. 1173–82).

Berberoğlu, E., and Uygun, S. (2012). Çevre Farkındalığı-Çevre Tutumu Arasındaki İlişkinin Yapısal Eşitlik Modeli İle Sınanması. *Uludağ Üniversitesi Eğitim Fakültesi Dergisi*, 459–73.

Berry, W. D., and Feldmen, S. (1985). *Multiple regression in practice (No. 50).* London: Sage.

Bhutto, M. H., Tariq, B., Khuwaja, F. M., Ahmed, K., Azhar, S., and Han, H. (2020). Predicting consumer purchase intention toward hybrid vehicles: testing the moderating role of price sensitivity. In *European Business Review.* Emerald.

Boztepe, A. (2012). Green Marketing and Its Impact on Consumer Buying Behavior. *European Journal of Economic and Political Studies*, 5–21.

Carlson, D. H. (2004). Environmental concern in South Africa: The development of a measurement scale. South Africa.

Chan, R. (2001). Determinants of Chinese Consumers' Green Purchase Behavior. In *Psychology and Marketing* (pp. 389–413). John Wiley & Sons.

Chang, M. (1998). Predicting Unethical Behavior: A Comparison of the Theory of Reasoned Action and the Theory of Planned Behavior. In *Journal of Business Ethics* (pp. 1825–34).

Chaudhary, R., and Bisai, S. (2018). Factors influencing green purchase behavior of millennials in India. In *Management of Environmental Quality: An International Journal* (pp. 798–812). Emerald.

Chen, A., and Peng, N. (2012). Green hotel knowledge and tourists' staying behavior. In *Annals of Tourism Research* (pp. 2211–19). Oxford: Elsevier.

Craney, T. A., and Surles, J. G. (2002). Model-Dependent Variance Inflation Factor Cutoff Values. In *Quality Engineering* (pp. 391–403).

Dean, M., Raats, M., and Stepherd, R. (2012). The Role of Self-Identity, Past Behavior, and Their Interaction in Predicting Intention to Purchase Fresh and Processed Organic Food. In *Journal of Applied Social Psychology* (pp. 669–88).

Dunlap, R., and Van Liere, K. (1978). The new environmental paradigm a proposed measuring instrument and. *The Journal of Environmental Education*, 10–19.

Dunlap, R., Liere, K., Mertig, A., and Jones, R. E. (2000). Measuring Endorsement of the New Ecological Paradigm: A Revised NEP Scale. In *Journal of Social Issues* (pp. 425–42).

Eagly, A., and Chaiken, S. (1993). *The psychology of attitudes.* Texas: Harcourt Brace Jovanovich College.

Elias, E., and Lin, C. K. (2014). The empirical study of green buildings (residential) implementation: perspective of house developers. *The 5th Sustainable Future for Human Security (SustaiN 2014)* (pp. 708–16). Elsevier.

Erdil, M. (2018). Understandıng the drivers of generation y consumers' green purchase intention: price sensitivity as a moderating variable. In *Journel of Business, Economics and Finance* (pp. 89–90). Press Academia.

Field, A. (2007). Discovering statistics using SPSS Second edition. Los Angeles: Sage Publication.

Foxall, G., and James, V. (2003). The behavioral ecology of brand choice: How and what do consumers maximize? In *Psychology of Marketing* (pp. 811–36). Wiley.

Global Alliance for Buildings and Construction. (2020). Retrieved on June 30, 2022, from https://globalabc.org/.

Goldsmith, R., Kim, D., Flynn, L., and Kim, W.-M. (2005). Price Sensitivity and Innovativeness for Fashion Among Korean Consumers. In *The Journal of Social Psychology* (pp. 501–08).

Hahnel, U., Ortmann, C., Korcaj, L., and Spada, H. (2014). What is green worth to you? Activating environmental values lowers price sensitivity towards electric vehicles. In *Journal of Environmental Psychology* (pp. 306–19). Elsevier.

Hair, J. F., Babin, B., Money, A. H., and Samouel, P. (2003). *Essentials of Business Research Methods 1st Edition.* United States of America: Wiley.

Hair, J., Anderson, R., Tatham, R., and Black, W. (1995). *Multivariate data analysis.* Prentice-Hall.

Hair, J., Black, W., Babin, B., and Anderson, R. (2010). Multivariate Data Analysis Seventh Edition. Hall International Inc.

Han, H., Hsu, L.-T., and Sheu, C. (2010). Application of the Theory of Planned Behavior to green hotel choice: Testing the effect of environmentally friendly activities. In *Tourism Management* (pp. 325–34). Elsevier.

Hanson, C. (2013). Environmental Concern, Attitude toward Green Corporate Practices, and Green Consumer Behavior in the United States and Canada. *ASBBS E-Journal, 9*(1), 62–70.

Haws, K., Winterich, K., and Naylor, R. (2014). Seeing the world through GREEN-tinted glasses: Green consumption values. *Journal of Consumer Psychology*, 336–54.

Hayes, D. (1990). The Green Decade. In *The Amicus Journal* (pp. 24–29).

Heffner, R., Kurani, K., and Turrentine, T. (2007). Symbolism in California's early market for hybrid electric vehicles. In T. R. 12. California: Elsevier.

Hsu, C.-L., Chang, C.-Y., and Yansritakul, C. (2017). Exploring purchase intention of green skincare products using the theory of planned behavior: Testing the moderating effects of country of origin and price sensitivity. In *Journal of Retailing and Consumer Services* (pp. 145–52). Elsevier.

"Sertifikalı projeler," last modified January 1, 2021. Retrieved from: https://cedbik.org/tr/sertifikali-projeler.

Huete Alcocer, N. (2017). A Literature Review of Word of Mouth and Electronic Word of Mouth: Implications for Consumer Behavior. In *Frontiers in Psychology Frontiers in Psychology (Vol.8)* (p. 1256).

Jaiswal, D., and Kant, R. (2018). Green purchasing behaviour: A conceptual framework and empirical investigation of Indian consumers. In *Journal of Retailing and Consumer Services* (pp. 60–69). Elsevier.

Kaiser, F., and Gutscher, H. (2003). The Proposition of a General Version of the Theory of Planned Behavior: Predicting Ecological Behavior. In *Journal of Applied Social Psychology* (pp. 586–603).

Kalafatis, S., Pollard, M., East, R., and Tsogas, M. (1999). Green marketing and Ajzen's theory of planned behaviour:a cross-market examination. In *Journal of Consumer Marketing* (pp. 441–60).

Kalaycı, Ş. (2010). *SPSS Uygulamalı Çok Değişkenli İstatistik Teknikleri*. Ankara: Asil Yayın Dağıtım.

Karatekin, K. (2014). Social Studies Pre-service Teachers' Awareness of Solid Waste and Recycling. In *Procedia–Social and Behavioral Sciences* (pp. 1797–1801). Elsevier.

Kats, G. (2003). Green Building Costs and Financial Benefits. Massachusetts Technology Collaborative.

Kaul, A., and Wittink, D. (1995). Empirical Generalizations about the Impact of Advertising on Price Sensitivity and Price. In *Marketing Science* (pp. 151–60). İnforms.

Khaleeli, M., and Jawabri, A. (2021). The effect of environmental awareness on consumers' attitudes and consumers' intention to purchase environmentally friendly products: Evidence from United Arab Emirates. In *Management Science Letters* (pp. 555–60).

Khare, A. (2015). Antecedents to green buying behaviour: a study on consumers in an emerging economy. In *Marketing Intelligence & Planning* (pp. 309–29). Emerald.

Kibert, C. (2004). Green Buildings: An Overview of Progress. *Florida State University Journal of Land Use and Environmental Law*, 491–501.

Kotchen, M., and Reiling, S. (2000). Environmental attitudes, motivations, and contingent valuation of nonuse values: a case study involving endangered species. In *Ecological Economics* (pp. 93–107). ELSEVİER.

Kubba, S. (2010). *LEED Practices, Certification and Accreditation Handbook*. Elsevier.

Kumar, B., K. Manrai, A., and Manrai, L. (2017). Purchasing behaviour for environmentally sustainable products: A conceptual framework and empirical study. In *Journal of Retailing and Consumer Services* (pp. 1–9). Elsevier.

Li, G., Li, W., Jin, Z., and Wang, Z. (2019). Influence of Environmental Concern and Knowledge on Households' Willingness to Purchase Energy-Efficient Appliances: A Case Study in Shanxi, China. *Sustainability*.

Lichtenstein, D., Ridgway, N., and Netemeyer, R. (1993). Price Perceptions and Consumer Shopping Behavior: A Field Study. In *Journal of Marketing Research* (pp. 234–45). American Marketing Association.

Liobikienė, G., Mandravickaitė, J., and Bernatonienė, J. (2016). Theory of planned behavior approach to understand the green purchasing behavior in the EU: A cross-cultural study. In *Ecological Economics* (pp. 38–46). Elsevier.

Luchs, M., Naylor, R., Irwin, J., and Raghunathan, R. (2010). The Sustainability Liability: Potential Negative Effects of Ethicality. *Journal of Marketing*, 18–31.

Maichum, K., Parichatnon, S., and Peng, K.-C. (2016). Application of the Extended Theory of Planned Behavior Model to Investigate Purchase Intention of Green Products among Thai Consumers. *Sustainability, 8,* 1077.

McDonald, S., Oates, C., Thyne, M., Timmis, A., Carlile, C., and Timmis, A. J. (2015). Flying in the face of environmental concern: why green consumers continue to fly. *Journal of Marketing Management*, 1503–28.

Moons, I., and Pelsmacker, P. (2012). Emotions as determinants of electric car usage intention. In *Journal of Marketing Management* (pp. 195–237).

Mostafa, M. (2007). A hierarchical analysis of the green consciousness of the Egyptian consumer. *Psychology & Marketing*.

Mostafa, M. (2009). Shades of green: A psychographic segmentation of the green consumer in Kuwait using self-organizing maps. In *Expert Systems with Applications* (pp. 11030–38). Elsevier.

Mostafa, M. M. (2007). Gender differences in Egyptian consumers' green purchase behaviour: the effects of environmental knowledge, concern and attitude. *International Journal of Consumer Studies*, 220–29.

Ng, M., Law, M., and Zhang, S. (2018). Predicting purchase intention of electric vehicles in Hong Kong. In *Australasian Marketing Journal* (pp. 272–80). Elsevier.

Ozaki, R., and Sevastyanova, K. (2011). Going hybrid: An analysis of consumer purchase motivations. In *Energy Policy* (pp. 2217–27). Elsevier.

Park, H. S. (2000). Relationships among attitudes and subjective norms: Testing the theory of reasoned action across cultures. In *Communication Studies* (pp. 162–75).

Paul, J., Modi, A., and Patel, J. (2016). Predicting green product consumption using theory of planned behavior and reasoned action. In *Journal of Retailing and Consumer Services* (pp. 123–34). Elsevier.

Peattie, K. (1995). *Environmental Marketing Management: Meeting the Green Challenge*. Londan: Pitman Publishing.

Preacher, K. J., and Hayes, A. F. (2004). SPSS and SAS procedures for estimating indirect effects in simple mediation models. *Behavior Research Methods, Instruments, & Computers, 36*(4), 717–31.

Ramayah, T., Lee, J. C., and Mohamad, O. (2010). Green product purchase intention: Some insights from a developing country. In *Resources, Conservation and Recycling* (pp. 1419–27). Elsevier.

Rashid, N., and Shaharudin, M. (2017). Customer's purchase intention for a green home. In *International Journal of Procurement Management* (pp. 1–19).

Retzlaff, R. (2009). Green Buildings and Building: A New Area of Interest for Planners. *Journal of Planning Literature*, 3–21.

Rosa, J., Garbarino, E., and Malter, A. (2006). Keeping the Body in Mind: The Influence of Body Esteem and Body Boundary Aberration on Consumer Beliefs and Purchase Intentions. In *Journal of Consumer Psychology* (pp. 79–91). Lawrence Erlbaum Associates.

Sang, Y.-N., and Bekhet, H. A. (2015). Modelling electric vehicle usage intentions: an empirical study in Malaysia. In *Journal of Cleaner Production* (pp. 75–83). Elsevier.

Schultz, P. (2000). New Environmental Theories: Empathizing With Nature: The Effects ofPerspective Taking on Concern for Environmental Issues. In *Social Issues* (pp. 391–406).

SGuin, C., Pelletier, L., and Hunsley, J. (1998). Toward a model of environmental activism. In *Environment and Behavior* (pp. 628–52).

Shah, S. H., Aziz, J., Jaffari, A., Waris, S., Ejaz, W., Fatima, M., and Sherazi, S. (2012). The Impact of Brands on Consumer Purchase Intentions. In *Asian Journal of Business Management*, *4*(2) (pp. 105–10). Maxwell Scientific Organization.

Sinha, I., and Batra, R. (1999). The effect of consumer price consciousness on private label purchase. In *International Journal of Research in Marketing* (pp. 237–51). Elsevier.

Sivadasan, R., Quoquab, F., Mohommad, J., and Basiruddin, R. (2020). Residential properties with green living concept: what drives consumers to buy? In *International Journal of Ethics and Systems*. Emerald.

Sparks, P., and Shepherd, R. (1992). Self-Identity and the Theory of Planned Behavior: Assesing the Role of Identification with "Green Consumerism. In *Social Psychology Quarterly* (pp. 388–99). American Sociological Association.

Spruyt, A., Hermans, D., Houwer, J., Vandekerckhove, J., and Eelen, P. (2007). On the predictive validity of indirect attitude measures:Prediction of consumer choice behavior on the basis of aVective priming in the picture–picture naming task. In *Journal of Experimental Social Psychology* (pp. 599–610). Elsevier.

Sreen, N., Purbey, S., and Sadarangania, P. (2018). Impact of culture, behavior and gender on green purchase intention. In *Journal of Retailing and Consumer Services* (pp. 177–89). Elsevier.

Stern, P. (1992). Psychological Dimensions of Global Environmental Change. In *Annual Review of Psychology* (pp. 269–302).

Suki, N. M. (2013). Green awareness effects on consumers' purchasing decision: some insights from Malaysia. *International Journal of Asia Pacific Studies*, 49–63.

Tan, T. H. (2013). Use of Structural Equation Modeling to Predict the Intention to Purchase Green and Sustainable Homes in Malaysia. In *Asian Social Science* (pp. 181–91). Canadian Center of Science and Education.

Tarkiainen, A., and Sundqvist, S. (2005). Subjective norms, attitudes and intentions of Finnish consumers in buying organic food. In *British Food Journal* (pp. 808–22). Emerald.

Taylor, S., and Todd, P. (1995). Decomposition and crossover effects in the theory of planned behavior: A study of consumer adoption intentions. In *International Journal of Research in Marketing* (pp. 137–55). Elsevier.

Teng, Y.-M., Wu, K.-S., and Liu, H.-H. (2013). Integrating Altruism and the Theory of Planned Behavior to Predict Patronage Intention of a Green Hotel. In *Journal of Hospitality and Tourism Research* (pp. 299–315).

Varshneya, G., K. Pandey, S., and Das, G. (2017). Impact of Social Influence and Green Consumption Values on Purchase Intention of Organic Clothing: A Study on Collectivist Developing Economy. In *Global Business Review* (pp. 478–93). Sage.

Wijayaningtyas, M., Handoko, F., and Hidayat, S. (2019). The millennials' perceived behavioural control on an ecofriendly house purchase intention. *Annual Conference of Science and Technology*. Indonesia: IOP.

(2018). *World Green Building Trends 2018*. DODGE.

Xu, X., Hua, Y., Wang, S., and Xu, G. (2020). Determinants of consumer's intention to purchase authentic green furniture. In *Resources, Conservation and Recycling* (p. 156). 104721: Elsevier.

Yadav, R., and Pathak, G. S. (2016). Young consumers' intention towards buying green products in a developing nation: Extending the theory of planned behavior. In *Cleaner Production* (pp. 732–39). Elsevier.

Yee, L. (1997). Effect of Collectivist Orientation and Ecological Attitude on Actual Environmental Commitment the Moderating Role of Consumer Demographics and Product Involvement. In *Journal of International Consumer Marketing* (pp. 31–53).

Yeung, S. P.-M. (2004). Teaching Approaches in Geography and Students' Environmental. *The Environmentalist*, 101–17.

Yılmaz, V., and Dalbudak, Z. İ. (2018). Aracı Değişken Etkisinin İncelenmesi: Yüksek Hızlı Tren İşletmeciliği Üzerine Bir Uygulama. *Int. Journal of Management Economics and Business, 14*(2), 517–34.

Zhang, L., Chen, L., Wu, Z., Zhang, S., and Song, H. (2018). Investigating Young Consumers' Purchasing Intention of Green Housing in China. In *Sustainable* (pp. 1–15).

Zhang, Y., Kang, J., and Jin, H. (2018, February 2). A Review of Green Building Development in China from the Perspective of Energy Saving. China.

Zimmer, M., Stafford, T., and Stafford, M. (1994). Green issues: Dimensions of environmental concern. In *Journal of Business Research* (pp. 63–74). Elsevier.

Zukin, S., and Maguire, J. (2004). *Consumers and Consumption*. 173–97: Annual Review of Sociology.

Sustainability and Transportation

Can Transportation Go Green? A Review

7

GÜL ESIN DELIPINAR

Introduction

Globalization has created an increasingly competitive environment. In a competitive world, it is important for businesses to gain an advantage over their competitors, increase their profitability and reduce their costs. Businesses that are open to innovation make themselves adapted to the globalizing world.

Today, businesses have focused on environmental issues, especially in order to maintain competitive advantage. As a result, businesses valued and produced environmentally friendly goods and services and green practices. Particularly, the use of unleaded gasoline in vehicles, the use of less energy consuming light bulbs and the use of recycled papers are some of the actions carried out for environmental awareness.

The effects of environmental pollution on human life have started to threaten future generations, making it obligatory for consumers and governments to take the necessary precautions. These obligations affect the businesses and supply chains. As a result of this impact, companies have to adapt its green perception to the supply chain in order to increase profitability and efficiency by making some changes.

Christea, Hummels, Puzzello, and Avetisyan (2013) mention the international transportation is the cause of two-thirds of trade-related emissions in US exports, 75 percent of the trade-related emissions such as transport equipment, electronic equipment, machinery, and manufactures come from transportation (Hu and Liu, 2020).

Literature Review on Sustainability

As thinking of sustainable impact in transportation, companies focus on fuels, air pollution, congestion and noise pollution. Christopher (2016) suggests the triple bottom line to evaluate the business decisions on key points of sustainability.

- Environment: Pollution, climate change, the depletion of scarce resources
- Economy: The effect on people's livelihoods and financial security; the profitability of business
- Society: The reduction of poverty, the improvement of working and living conditions etc.

The three P's of people—people, profit, and planet—are interrelated and are important for companies to be sustainable and viable.

In the vision of Abu Dhabi Economic Vision 2030, the main focus is reducing the dependence of oil and its fluctuation to create a sustainable economy (Alameeri, Ajmal, Hussain, and Helo, 2017)

According to an approach, the necessary steps in sustainability are: reuse, remanufacturing, reconditioning, and recycling (Coyle, Langley Jr., Novack, and Gibson, 2017). In reusing, parts of a product can be reused after cleaning, checking and repair or individual parts. Remanufacturing has a meaning of "good as new." Auto parts, tires, and electronics can be remanufactured. Other term reconditioning brings the product in a working condition, but not "good as new" condition. Lastly, recycling is secondary use of materials. Glass bottles, cans, bottle, tires are recycled products in daily life.

The Kyoto Protocol is an international treaty to reduce Greenhouse Gas Emissions from aviation and shipping. The developed countries are mostly responsible from the Kyoto Protocol and they have a responsibility to adopt policies and report periodically. The transport sector is the largest consumer of fuel in the world. It consumes 49.5 million barrels per day (mb/d) of crude oil in 2014 and it constitutes 54.6 percent of world use. In 2015, the transport sector constitutes 23 percent of all carbon emissions. That is seen a threat for global warming (Stead, Wadud, Nash, and Li, 2019). According to Larson, Elias, and Viafara (2013), it is possible to reduce emissions and fuel consumption, however companies need to have a capital investment. Additionally, cooperation of stakeholders is necessary to move forward.

There are two approaches to calculate CO_2 emissions from road freight operations.

1. Fuel-Based: calculated based on fuel receipts includes the details about the type of fuel and quantity purchased or direct measurements of fuel use.
2. Activity-Based: calculated according to the activity level on vehicles. (McKinnon, Cullinane, Browne, and Whiteing, 2010).

The Kyoto Protocol has a period of applicability for the years 2008 to 2012 (first obligation period) and 2013 to 2020 (second obligation period). In order to keep climate protection process after 2020, the new agreement (Paris Agreement) has been active since 2015. Approved countries have a reduction emission targets Climate Action Tracker (2021). Countries involved to this agreement supposed to follow rules of the agreement; whereas countries not involved has to pay carbon tax to the European Union countries to stay in the global business. In 2016, 175 countries signed the agreement (Tong 2021).

For supply chain managers, the reasons for greenhouse gases are from industrial activities, such as manufacturing, energy production, and transportation. Because of globalization, products need to move a far place, which is a cause of the carbon footprint (Christopher, 2016).

The greenhouse gases include carbon dioxide, methane, nitrous oxide, and various fluorocarbons. These emissions form carbon footprints (Christopher, 2016). "Carbon footprint" is another term declared in green life. Although the definition is clear in which a certain amount of gaseous emissions related to climate change linked to human production and consumption activities, how to measure or quantify the carbon footprint is still debatable.

Traffic congestion is another reason of fuel consumption and CO_2 emissions (McKinnon, Cullinane, Browne, and Whiteing, 2010). It also causes less reliable delivery times, longer commutes, increasing fuel consumption and eventually more emissions (Larson, Elias, and Viafara, 2013). Christopher (2016) mentions that traffic congestion is related to city's infrastructure. In global trade, with the removal of trade barriers, the flow of products has increased. The sizes of containers are also increased in parallel direction. When the containers are unloaded or delays at port, it leads to traffic congestion. Additionally, Just in Time production also causes to environmental issues. Smaller but frequent movements have increased shipment and movements.

According to Larson, Elias, and Viafara (2013), traffic congestion problems can be solved with the help of investments in infrastructure, such as with tolling and truck routes that don't have congested areas. Otherwise it has costs such as, maintenance costs, more accidents, and reduced freight movement.

In measuring greenhouse emissions, Greenhouse Gas (GHG) Protocol has three scope levels. The first scope mentions the emissions from GHG sources that are owned or controlled by the company, direct emissions. The second scope has a view of indirect sources, such as electricity, heat, steam, and cooling. The last scope has a view of indirect emissions from production of purchased materials, outsourced activities, waste disposal and employee travel (Chopra and Meindl 2013).

According to McKinnon, Cullinane, Browne, and Whiteing (2010) freight transport composes 8 percent of energy-related CO_2 emissions worldwide. In the road transport, energy used in freight transportation is higher than cars and buses. In 2050, it is estimated that only shipping might take 15–30 percent of CO_2 emissions.

Fuel prices have an important impact on transportation efficiency. It especially affects cost and service and eventually affects shippers. Shippers create distribution centers to avoid fuel price volatility and price increases. Another way to decrease fuel consumption might be to use vehicle-routing algorithms to minimize the fuel consumption or carbon emissions of a fleet of vehicles. The fuel consumption in the night delivery strategies to reduce fuel consumption or cost can be another solution to decrease fuel consumption.

There were times the higher fuel prices reduced the high fuel consumption. Fuel price volatility, price changes have always been the center of attention for transportation carriers and shippers. Fuel efficiency reduces the carbon footprint of commercial transportation. Fleet managers are investigating to clean fuels, reduce miles traveled, consolidating load to obtain fuel efficiency.

Mentioning the green supply chain, there are other green strategies used in transportation for sustainability. "Don't ship air" or "Don't ship water" strategies take attention. The first phrase means empty space in a motor carrier trailer or railcar or less than capacity product sent is wasted, and costly. The main aim of the transportation is to reduce or minimize the space in transportation equipment (Novack, Gibson, Suziki, and Coyle, 2018). Chopra and Meindl (2013) support efficient loading of freight might increase fuel efficiency. This eventually cuts costs and gives less damage to environment.

The latter phrase, "Don't ship water," means not wasting water and reducing the transportation costs moving water in relatively short distances. Another example might be liquid detergent, which has a smaller plastic bottle, but has the same washing power (Novack, Gibson, Suziki, and Coyle, 2018). At this point, product design is essential in reducing transportation costs. IKEA, for instance, is selling disassembled products to take advantage of less volume of products, which eventually reduces emissions and energy use.

According to Chopra and Meindl (2013) from an environmental perspective; all firms should measure and report on these four categories:

1. Energy Consumption
2. Water Consumption
3. Greenhouse Gas Emissions
4. Waste Generation

Sustainability measures and focus change and depend on the companies' sector. For instance, pharmaceutical companies value waste management and water consumption (McKinnon, Cullinane, Browne, and Whiteing, 2010) while transportation companies report the sustainability company results on greenhouse emissions, fuel consumption, and transportation efficiency. (Chopra and Meindl 2013). The remaining parts of this chapter evaluate sustainability in different modes of transportation according to greenhouse emissions, fuel consumption, and transportation efficiency.

Sustainability in Different Modes of Transportation

Motor Carriers

The motor carrier industry has an essential role in the US economy during the twentieth century. However, it was not prevalent until World War I. After World War II, the railroad industry was in a competition with railroad industry. Because highways developed from the 1950s to 1991, they started to take the place of railways. Motor carriers eventually became the largest share of US freight movements compared to other modes of transportation (Novack, Gibson, Suziki, and Coyle, 2018).

Road transport is prevalent in terms of volume in other transport modes. According to 2017 data, freight transport by highways is 50 percent of total freight volume (Hu and Liu, 2020).Therefore, it is essential to mention first.

Motor carriers use fuel at high levels therefore every change in prices affect the fuel consumption in motor carrier industry. In 2008, diesel fuel expenses in motor carrier industry were three times higher than airlines (Novack, Gibson, Suziki, and Coyle, 2018).

According to Larson, Elias, and Viafara (2013), freight transport emissions are 95 percent carbon dioxide (CO_2). Trucks use fifty-four billion gallons of fuel each year in the United States. This amount represents 20 percent of air contaminant.

Non-fossil fuel use and bio-fuels might be another alternative to be efficient in transportation. Additionally, alternative fuel vehicles that work with compressed natural gas (CNG), ethanol (E85), liquefied petroleum gas (LPG), methanol (M85) or even electricity might be preferred to be sustainable (Caamaño, Olmedo, García, and Prado, 2018).

Biofuel is a fuel that is both a by-product of and a product that can easily be produced in a natural environment. It does not affect current crop production and does not provide any harmful effect when used. Examples for the first-generation fuels would be based on sugar, starch, or vegetable oil and hence comprise fuels like biodiesel, bio-alcohol, and bio-gas, among others. The second-generation bio-fuels, on the other hand, would be made from sustainable feedstock and comprise fuels like wood diesel, bio-methanol, bio-hydrogen, algae-fuel, and so on (Melis, 2014, p. 524)

Zhang, Song, Yang, and Li (2020) suggest the technologies, such as satellite communication systems, can help companies to be sustainable and increase transportation efficiency. Technologies mentioned are vehicle routing technologies in delivering products, electrification alternative drives, and autonomous optimization using artificial intelligence.

Raslavičius, Keršys, Starevičius, Sapragonas, and Bazaras (2014) highlight that electric vehicles have 53 percent less emissions than conventional fuel-powered vehicles, however conventional fuel vehicles are more preferable due to low investment costs. Larson, Elias, and Viafara (2013) propose green driver training, and driver skills, vehicle utilization, and routing and intelligence transportation systems are essential for fuel consumption and transportation efficiency. Goel (2009), Caamaño, Olmedo, García, and Prado (2018), and Larson, Elias, and Viafara (2013) all state that in order to provide sustainable transport and energy efficiency is to prefer alternative transport modes, especially rail, or water carriers.

Air Carriers

Wilbur and Orville had their first flight in 1903 and the federal government bought their invention, which was the first spark of air transportation. In 1908, the government used it for the US Post Office, followed by passenger transportation accompanied with the mail business. Later, it was noted that when the time is limited and valuable, air carriers are faster, have a larger volume of transportation and are more fuel efficient (Novack, Gibson, Suziki, and Coyle, 2018).

According to Bartle, Lutte, and Leuenberger (2021), air cargo has an essential role in the transportation of goods in global trade. Air carriers have a capacity of goods worth of 6.8 trillion USD every year. This value is the 35 percent of global trade. Walala and Mutinda (2013) mention that the aviation industry causes pollution, and bigger and older airplanes cause more pollution. Air carriers are also intensive users of fuel. If fuel costs decrease, it will bring more freight to the airlines (Novack, Gibson, Suziki, and Coyle, 2018). It is efficient and sustainable if the aviation sector uses biofuels. Biofuels are sustainable fuel, and it might be the "only real good option to reduce aviation emissions by 2050" (Bartle, Lutte, and Leuenberger, 2021). Emissions are 19 percent of total aviation global emissions. Eleven percent of that is from belly cargo, and 8 percent is from freighters (Bartle, Lutte, and Leuenberger, 2021).

Water Carriers

Water transportation involves inland and domestic waterways, and it has been an important connection for freight and people movement for years. In Europe and other countries, rivers have a great value in connecting cities and countries for transportation. The waterways were efficient types of transportation before the railroads (Novack, Gibson, Suziki, and Coyle, 2018).

Water carriers are very fuel efficient and they are not sensitive to fuel price changes (Novack, Gibson, Suziki, and Coyle, 2018). Lee, Kwon, and Ruan (2019) mention that stakeholders focus on less sustainability issues in the maritime sector. This is because water carriers are far from center of city life and they have less interaction with cities.

According to the OECD in 2010, a projection for 2020 maritime fuel consumption was 30 percent. For 2050 CO_2 emissions are expected to be two or three times that of current levels. Therefore, in 2050 it is expected that fossil fuel use is lowered as much as possible so that it will create less air pollution.

Rail Carriers

Rail carriers were the dominant mode of transportation for both passengers and goods freight during the first decades of twentieth century.

However, with the government-funded roads, and superior service motor carriers and automobiles, inland water transportation and air transportation lost their prevalence to other transportation modes. It also had difficulty to move goods to short distances, which was motivation for other modes of transportation. However, it is still important for the economy with its high revenues and number of employed people (Novack, Gibson, Suziki, and Coyle, 2018).

Railways are the most environmentally friendly mode of transportation (Raslavičius, Keršys, Starevičius, Sapragonas, and Bazaras, 2014), and also a cost effective mode of transportation in carrying goods. Compared to road transportation, rail carriers are more efficient; however, it has fixed routes and transshipments are accomplished at terminals (Zhao, 2020). Reddy, Anand, and D'Sa (2000) mention that their scope of study is limited to road and rail transport. This is because air and water have small fractions of demand.

The rail carriages have an advantage in fuel efficiency, and do not have a cost pressure as other modes of transportation do (Novack, Gibson, Suziki, and Coyle, 2018). According to Alvarez, Martinez, and Franco (2013), the result of their study shows that fuel consumption and emissions in railway modes have a greater change compared to road modes. Additionally, rail modes have fuel consumption and emission levels depending on the source of energy used the type of freight carried, and the route profile.

Fuel efficiency can be maintained by using lighter wagons, low friction bearings, and on-board lubricators that provide a 25 percent reduction in fuel consumption. Additionally, technology and deregulations in tariffs can provide efficiency in fuel (Reddy, Anand, and D'Sa, 2000). Stead, Wadud, Nash, and Li (2019) suggest that biodiesels are low-carbon and are alternative fuel sharing similar chemical properties to diesel. It can reduce carbon emission and it is also good for human health. However, it has higher costs compared to fossil fuels. These costs are recommended to be paid by the consumer or the taxpayer. Rail carriers are more suitable than motor carriers to use biodiesels since there is a need of fewer stations.

Palumbo (2013) states that although transportation constitutes a high percentage of carbon emissions, only 2 percent of this amount of emissions is caused by rail. Because rail transport is seen as the sustainable transportation, the European Union is focusing on railways to serve European citizens. Therefore, the train has a better performance in terms of energy consumption than cars and planes in the transport sector. Electrical trains can be used for renewable energy; however, it is not completely emission-free. They give out particles from rails, brakes, wheels, and carbon contract strips. Another issue with the railways is the noise in urban areas (Behrends, 2012).

Conclusion

This is a review of modes of transportation in terms of sustainability. Transportation has the highest rate of carbon emissions. However, some of the transportation modes have less carbon emission and they are more environmentally friendly. Eco-friendly modes of transportation might be a preference, such as railways or water carriers. However, because they take more time for products to arrive, motorways or airways might be preferable. Additionally, there might not be appropriate railways or ports. Therefore, it becomes a necessity to use motorways or airways instead.

Considering biodiesels as a fuel type, they can be used to lower carbon emissions. Biodiesels are considered as a type of renewable energy. Other energy sources are wind, solar, hydro, tidal, and geothermal energy. Different energy types might also be used for different sources of transportation. Another solution to lower emissions might be using electrical vehicles.

Electrical vehicles do not only have low emissions, but also have a higher efficiency. However, it is important how the energy is generated. If it is generated from coal, then the efficiency is less than gas powered stations. Electricity stored in batteries might need more time and infrastructure to be used. Additionally, batteries have lower energy density compared to fossil fuels. Batteries also need high lithium-ion materials which require cobalt. Politicians and car manufacturers are planning to increase the nickel percentage although nickel deposits are being depleted. It is thought that the industry will be more dependent on lateritic ore and batteries still need more time to develop, so the industry will still be using diesel vehicles in the near future (Dizard, 2018).

Cost-effective or not, protecting the sustainability of the environment will become mandatory with the Paris Agreement, which means that countries will be forced to pay carbon taxes to stay in business. The future studies might involve the different modes of transportation and how to lower carbon emissions through 2050. Because there is also pollution as a result of transportation, it will be no longer a preference to use alternative fuels or electrical vehicles. Additionally, the conventional way of living will be more costly with the new laws and regulations. Transportation can become greener, if new actions are taken by companies and if countries have an intention to follow the "green" actions mentioned above.

References

Alvarez, A., Martinez, P. P., and Franco, I. (2013). Energy Consumption and Carbon Dioxide Emissions in Rail and Road Freight Transport in Spain: A Case Study of Car Carriers and Bulk Petrochemicals. *Journal of intelligent transportation systems*, 233–44.

Alameeri, A., Ajmal, M. M., Hussain, M., and Helo, P. T. (2017). Sustainability practices in the aviation sector: a study of UAE-based airlines. *International Journal of Sustainable Society*, *9*(2), 119–47.

Bartle, J., Lutte, R., and Leuenberger, D. (2021). Sustainability and Air Freight Transportation: Lessons from the Global Pandemic. *Sustainability*. *13*(7), 37–38.

Behrends, S. (2012). The significance of the urban context for the sustainability performance of intermodal road-rail transport. *Social and Behavioral Sciences*, 375–86.

Caamaño, P., Olmedo, B., García, R. M., and Prado, P. J. (2018). How to Do Road Transport More Sustainable? A Literature Review. *Springer International Publishing*.

Christopher, M. (2016). *Logistics and Supply Chain Management*. Pearson Education.

Chopra, S., and Meindl, P. (2013). *Supply chain management*. Essex, England: Pearson Education.

Cristea, A., Hummels, D., Puzzello, L., and Avetisyan, M. (2013). Trade and the greenhouse gas emissions from international freight transport. *Journal of environmental economics and management, 65*(1), 153–73.

Coyle, J. J., Langley Jr., C. J., Novack, R., and Gibson, B. J. (2017). *Supply Chain Management: A Logistics Perspective*. Cengage Learning.

Climate Action Tracker (2021). Global Climate Agreements. Retrieved on June 30, 2022 from: https://www.cfr.org/backgrounder/paris-global-climate-change-agreements.

Goel, A. (2009). A Roadmap for Sustainable Freight Transport. Retreived from: file:///Users/natalya1/Downloads/Go09_MMDTA.pdf.

Dizard, J (2018). Going green is not that easy for electric vehicles. https://www.ft.com/content/daabb4ad-5766-3c4b-86ca-2b9c92da4c49.

Hu, Y., and Liu, D. (2020). A calculation framework and tools to estimate freight rate and carbon emissions for road transport A study at Volvo Group Trucks Operations. Sweden.

Larson, P., Elias, A., and Viafara, J. (2013). Toward sustainable trucking: Reducing emissions and fuel consumption. *Transportation Journal*, 108–20.

Lee, P. T.-W., Kwon, O. K., and Ruan, X. (2019). Sustainability Challenges in Maritime Transport and Logistics Industry and Its Way Ahead. *Sustainability*, 11, 1331.

McKinnon, A., Cullinane, S., Browne, M., and Whiteing, A. (2010). *Green Logistics Improving the Environmental Sustainability of Logistics*. Kogan Page Limited.

Melis, W. J. (2014). The future of electrical vehicles. *Autonomous Vehicles: Intelligent Transport Systems and Smart Technologies*, 509–30.

Novack, Robert A.; Gibson, Brian J.; Suzuki, Yoshinori; and Coyle, John J., "Transportation: A Global Supply Chain Perspective" (2018). *Supply Chain Management Books.*

OECD (2010) Globalization, Transport and the Environment, www.oecd.org /env/trans.

Palumbo, M. (2013). Environment Sustainability of Rail Transportation. *Railwaysignaling.eu*, 1–4. Retreived from: https://www.railwaysignalling.eu /wp-content/uploads/2014/06/Environmental_Sustainability_of_Rail_trans portation__.pdf.

Raslavičius, L., Keršys, A., Starevičius, M., Sapragonas, J., and Bazaras, Ž. (2014). Biofuels, sustainability and the transport sector in Lithuania. *Reneable and Sustainable Energy Reviews*, 328–46.

Reddy, A., Anand, Y., and D'Sa, A. (2000). Energy for a sustainable road/rail transport system in India. *Energy for Sustainable Development*, 4(1), 29–44.

Stead, C., Wadud, Z., Nash, C., and Li, H. (2019). Introduction of Biodiesel to Rail Transport: Lessons from the Road Sector. *Sustainability*, 11, 904.

Tong, S (2021). How the EU's Carbon Border Tax will work. https://www .wbur.org/hereandnow/2021/08/10/carbon-border-tax-eu.

Zhang, H., Song, M., Yang, X., and Li, P. (2020). What are Important Technologies for Sustainable Development in the Trucking Industries of Emerging Markets? Differences between Organizational and Individual Buyers. *Sustainability.*

Zhao, M. (2020). Sustainability of Rail Transport in Africa: A Case Study of Kenya's Standard Gauge Railway. *Journal of Environmental Science and Engineering*, 17–27.

How Much Energy Sector in Turkey is Circular? 8

NATALYA KETENCI

Introduction

It is difficult to identify exactly when the notion of the Circular Economy appeared; there are various materials and evidence that exist in the literature. The circular economy only started to be widely discussed in 1990 with the work of Pearce and Turner (1990), where authors for the first time presented the concept of the circular economy, which contradicted the traditional linear system of economic growth. The circular economy proposes that the waste of different stages of a product's life turn to be input for the production.

The Circularity Gap Report (CGR) was launched in 2018. According to its statistics, the world economy was 9.1 percent circular in 2018, however, according to CGR (2022) this rate falls to 8.6 percent, which could occur because of the COVID-19 pandemic. There is a growing number of activities to make a circular economy more widespread and to find solutions for an efficient circularity. For example, the Circular Economy and Material Value Chains project is undertaken by the World Economic Forum and is directed on an increase of collective actions, of public and private sector leaders to make the circular economy notion more widespread in the world and to bring these actions to the life.

The European Commission developed and adopted the Circular Economy Action Plan in 2015. In March 2020, the program was updated to the New Circular Economy Action Plan (COM, 2020) to decrease waste and increase productivity. The plan reviews the key product value chains with the highest share of waste and significant potential for recycling, such as electronics and ICT, batteries and vehicles, packaging, plastics, textiles, construction and buildings, food, water, and nutrients. The plan proposes thirty-five actions to decrease waste and to create conditions for circularity in other possible areas of an economy.

Since 2016, the Business Council for Sustainable Development Turkey and European Bank for Reconstruction and Development have been developing a platform for an awareness increasing and preparing Turkey for a transition to the circular economy. The Turkey Material Marketplace was one of the first online platforms that allow companies to reuse materials of different firms, it allows inter-industrial cooperation in the context of the circular economy. Lately, this platform was renamed the Circular Economy Platform Turkey, where it was extended with technical assistance, knowledge, and training portals to make an efficient transition to the circular economy.

Among OECD countries, South Korea has the highest ratio of municipal waste recycling, 56.5 percent. The countries with the lowest ratios of municipal waste recycling are Israel, 6.8 percent, and Turkey, 11.9 percent. This ratio composes of 32.1 percent and 48 percent for the United States and the EU, respectively, OECD (2021). The landfill is the oldest way of waste disposal, which presents the burying the waste. After the second world war, new technologies were innovated to make it more efficient by using less space, however, still, it contradicts the circular economy and damages the environment. According to the OECD statistics, the highest ratio of municipal waste that is going to landfills is in Turkey, Greece, and Israel with 87.7 percent, 77.7 percent, and 76.5 percent, respectively. At the same time Belgium, Denmark, Finland, Germany, Japan, Sweden, and Switzerland have almost no waste sent to landfills.

In 2017, Zero Waste Project was initiated in Turkey to decrease the volume of non-recyclable waste and increase the recycling rate to 35 percent. In 2021 the project received a UNDP "Sustainable Development Goals (Global Goals) Action Award." This project is highly beneficial for municipal waste recycling (Erat and Telli, 2020).

This chapter reviews the most polluting sectors, that require to be involved in the circular economy at a higher speed. This review is followed by an analysis of the energy sector in Turkey and how the green energy alternatives are actually environmentally friendly. The next part overviews sectors that have the highest potential for circularity, being the most polluting sectors, and their conditions in Turkey. The following part analyses the energy sector in Turkey and its transition way to the circular economy.

The Most Polluting Industries with a High Potential for the Circular Economy

Packaging Industry

Some sectors have more potential for the circular economy compared to others. For example, products from industries like paper, plastic, glass, and metals are relatively widely involved in the circular economy. These materials are the main parts of the packaging industry, which every year create millions of tons of waste all over the world. In 2019 packaging sector in the EU generated 32.3 million tons of waste, Eurostat (2020), while the United States generated 82.8 million tons, which makes 28.1 percent of municipal waste. The lack of a recycling strategy in an individual country may lead to a global catastrophe, thus China, Indonesia, Philippines, Thailand, and Vietnam generate more than half of the world's plastic waste leakage into the ocean (Conservancy, 2017), which made these countries as focus countries for international environmental organizations.

Most of the literature on the packaging industry's involvement in the circular economy is related to plastic packaging, as it is one of the environment's damaging materials. Plastics poison water, soil, and air with toxins that kill wildlife around us, and most plastics are lasts for a longer time than our lives, continuing to damage the environment for centuries after their short usage.

Textile Industry

The textile industry is the second polluting industry after the oil industry, being one of the biggest global industries due to its fast-changing fashion (Malik, et al. 2014). In the last two decades, the textile industry production and its average annual consumption in the world doubled, where more than 85 percent ends up in landfills and only 15 percent is recycled (Shirvanimoghaddam et al., 2020). The enormous amount of waste is worsened by toxic substances that are used during the production of textile products, particularly during dyeing and color printing processes while spinning weaving and knitting are highly dependent on fossil fuel energy (Roos, et al., 2015). Another important problem in the textile industry is the underutilization of clothing. In 2014, global average consumption of clothing products increased by 60 percent but the utilization decreased twice, people keep and wear clothes twice less than in 2000 (McFall-Johnsen, 2020). The textile industry is involved in the pollution of all organs responsible for the life of our planet, water, soil, and air. Thus, 10 percent of air pollution is coming from the textile industry, it is more carbon emissions than the total from all international flights and sea transportation.

Oil Industry

The oil and gas industry is the most polluting industry in the world. It is polluting not only areas of raw materials extraction, where water and soil can be contaminated by wastes, but this industry is becoming a reason for the environmental degradation due to its products and byproducts. The main petroleum products are various fuels for all types of transportation, fuel oils for heating, asphalt, plastic, wax, and many others. By-products or products derived from petroleum are all products around us. Almost all products that are used in industrial production and in individual consumption are derived to a different extent from petroleum products, for example, tires, eyeglasses, ballpoint pens, clothing, basketballs, nail polish, lipstick, soap, shoes, medicine, and many others. Waste of all these products emits a variety of pollutants that damage not only the environment but the wild world as well. The harm of the petroleum sector to our health is one of the matters that is widely discussed to bring alternatives, however, the focus of this book is a decrease of new resources usage to a zero-waste life. Together with disadvantages, the oil industry has many advantages compared to its alternative for everyday life consumption and production processes. Therefore, an increase in exposure of the oil industry to a circular economy will not only make the environment cleaner but will allow making oil resources available for our future generations for a longer period.

Energy Sector in Turkey

The energy sector includes companies that produce or provide energy power. Energy power production takes place using non-renewable, such as oil, gas, coal, or renewable energy resources like biomass, solar, wind, geothermal, and hydro powers. Together with the harmful impact of non-renewable energy resources waste on our environments, the usage of these resources cannot last forever due to their scarcity. The lack of efficient alternatives undermines the sustainable development and growth of world production. Renewable energy sources are alternative energy sources that do not depend on stocks and are more environmentally friendly, however currently are less efficient compared to fossil fuels.

Turkey is highly dependent on non-renewable energy sources, where 72 percent of greenhouse gas emissions are energy sector-related, with a transport sector emission of 21 percent (Cevheribucak, 2021). The high dependence of Turkey on energy imports makes sustainable development more difficult with an increasing necessity for recycling technologies. Solar energy in Turkey has the highest potential as an alternative to non-renewable energy and is forecasted to satisfy 43.2 percent of the total electricity demand by 2050 (Kilickaplan et al. 2017).

Solar energy is a domestic energy power source with relatively low greenhouse gas emissions (production, transportation, and installation processes of solar panels are involved in the energy consumption generating GHG). However, solar energy similar to other renewable energy sources is not fully green. Renewable energy sources have wastes as well as other industries that may be harmful and contaminate an ecosystem.

The average life of a solar panel is between twenty-five and thirty years; after the 2000s, installation of solar panels all over the world increased significantly. However, there is not yet enough solid technology for a solar panels recycling, which is important in terms of scarce and valuable materials used in its production, such as aluminum, glass, silicone, and silver. The growing waste of used and not recycled solar panels increases the contamination of the environment with toxic materials. The waste of solar panels per unit of energy is 300 times more than the waste of nuclear power plants (Desai and Nelson, 2017). The technology for solar panel recycling is under research process all over the world including Turkey. Mersin University in Turkey became one of the contributors to the Sunrise project titled "Solar Energy for a Circular Economy," selected by the European Commission, and funded by the Horizon 2020 program (Abbott, 2019). The aim of this project is to research and prepare technologies that will make solar power production an efficient and recyclable alternative to fossil fuels.

The industrial production and transportation sector in Turkey is highly dependent on imported fuels, to ensure continuous and sufficient growth of the economy it is essential to decrease the dependence of these sectors on fossil fuels. One of the solutions for the transportation sector is the usage of electric vehicles. There is an increase in the global tendency for the use of electric vehicles, however not very rapid due to required substantial investments in the infrastructure, such as charging stations. In 2019 China's sales of electric vehicles composed more than half of world sales, of its high population, followed by the US, Norway, Germany, and France (Gönül et al. 2021). At the global level still there is less than 1 million sales and it is a very small ratio of total sales of vehicles.

In 2021 Turkey was the fourth-largest producer of vehicles in the EU, after Germany, Spain, and France. However, electric cars in Turkey are not popular. Hybrid-type cars are more highly demanded compared to other types of electric vehicles (battery, plug-in hybrid). Sales of hybrid cars compose around 97 percent of total electric car sales. The preference for fuel or hybrid automobiles is explained by the lack of infrastructure in Turkey that would support the charging process. The charging station is mainly located in a few big cities in Turkey, which eliminate the possibility to travel with electric vehicles to another city. However, in the January–March period of 2022, sales of electric vehicles increased by 243.9 percent to 1,073 units compared to the same period of the previous year, while sales of hybrid vehicles decreased by 17.1 percent to 11,227 units (ODD, 2022).

The usage of electric vehicles significantly decreases carbon emission, however, there is another waste problem that appears, which is an increasing number of toxic batteries. The number of hybrid or electric automobiles is growing every day worldwide, together with the number of electric bikes and scooters. These transport vehicles are safe for the environment with zero carbon emissions during their use, however, after the life span, the waste of the batteries used is becoming harmful and dangerous. A decrease in the use of carbon-emitting energy sources does not improve the environment if the circular economy is not involved in the production process.

Conclusion

This chapter overviews the most polluting sectors, such as packaging, textile, and oil sectors, and discusses the need for more efficient involvement of these sectors in the circular economy. Escalating the population leads to continuously increasing waste all over the world. Lack of proper waste management allows toxic materials to contaminate the environment and wildlife. One of the polluting industries in the world is the energy sector, which pollutes the environment not only in areas of extraction but all over the world using its products and derived products, which exist in all areas of industrial production and consumption. The energy sector in Turkey is characterized by a high dependency on foreign energy sources, which makes economic growth more difficult due to the continuously increasing energy costs. Therefore, Turkey, like Europe plans to improve the renewable energy infrastructure and decrease its dependency on energy imports. The most prospective alternative for non-renewable energy in Turkey is solar energy. With advances in technologies, solar energy may become an efficient alternative to natural gas in electricity generation and water heating, with no carbon emissions. However, the usage of solar panels, which are made of scarce and toxic materials, makes this sector even more damaging to the environment. Non-recycled solar panels generate 300 times more waste than nuclear power stations. There are various funds in the world that support research on technologies to involve the solar energy sector in the circular economy and to make it green. Turkey is involved in the research on circularity for the energy sector, however, it is a long process to go.

Another energy sector that is growing worldwide to make the transition to a green economy is electric vehicle usage, which significantly reduces carbon emissions. At the same time, a growing number of non-recyclable batteries in the waste produces toxic elements that pollute and poison the environment. Turkey is behind many developed and European countries in terms of electric automobile usage yet, because of a lack of the infrastructure, however other electric vehicles like scooters and bikes do not underestimate the problem of growing toxic batteries.

Transition to the green economy in Turkey, like other countries in the world, does not pass smoothly, reduction in carbon emission problems leads to new problems like a growing toxication of the environment. Therefore, continuous, and advanced research is required in order for the transition to a green economy to become as clean as possible.

References

Abbott A. (2019). Six teams vie for billion-euro funding. Nature 566, 164–65.

CGR (2022) The Circularity Gap Report 2022 (pp. 1–64, Rep). Amsterdam: Circular Economy.

Cevheribucak, G. (2021) Energy Transition and Sustainable Road Transportation in Turkey: Multiple Policy Challenges for Inclusive Change. Energy Justice in the Era of Green Transitions, 12.07.2021.

COM (2020) A new Circular Economy Action Plan: For a cleaner and more competitive Europe. European Commission, 98, Brussel, 11.3.2020.

Conservancy (2017) Stemming the Tide: Land-based strategies for a plastic-free ocean. McKinsey Center for Business and Environment. Full report, 04.

Desai, J. and M. Nelson (2017). Are we headed for a solar waste crisis? Environmental Progress, 21/6/2017. Retrieved from: https://conspiracytech.com/Are%20we%20headed%20for%20a%20solar%20waste%20crisis_%20%E2%80%94%20Environmental%20Progress.pdf.

Erat, S. and A. Telli (2020) Within the global circular economy: A special case of Turkey towards energy transition. MRS Energy and Sustainability, 7, Article number 24.

Eurostat, 2020. https://ec.europa.eu/eurostat/statistics-explained/index.php?title=Packaging_waste_statistics.

Gönül, Ö., Duman, A. C. and Ö. Güler (2021) Electric vehicles and charging infrastructure in Turkey: An overview. Renewable and Sustainable Energy Reviews, 143, 110913.

Kilickaplan A., Bogdanov D., Peker O., Caldera U., Aghahosseini A., and Breyer C. (2017). An energy transition pathway for Turkey to achieve 100% renewable energy powered electricity, desalination and non-energetic industrial gas demand sectors by 2050. Solar Energy 158, 218–35.

Malik, A., Akhtar, R., and Grohmann, E. (2014). Environmental deterioration and human health: Natural and anthropogenic determinants. Environmental Deterioration and Human Health: Natural and Anthropogenic Determinants, 1–421.

McFall-Johnsen, M. (2020) These facts show how unsustainable the fashion industry is. World Economic Forum. Retrieved from: https://www.weforum.org/agenda/2020/01/fashion-industry-carbon-unsustainable-environment-pollution/.

Production statistics OICA (2021) (International Organization for Motor Vehicles Manufacture). https://www.oica.net/category/production-statistics/2021-statistics/.

OECD (2021). Environment at a Glance 2020: OECD Indicators—Circular economy—waste and materials. OECD Publishing. May 22, 2021.

ODD (2022) Otomotiv Distribütörleri Derneği, Basın Bülteni, 04.04.2022. Retrieved from: https://www.odd.org.tr/folders/2837/categorial1docs/3141/ODD%20Bas%c4%b1n%20Bulteni%204%20Nisan%202022.pdf.

Pearce, D. W. and R. K. Turner (1990), Economics of Natural Resources and the Environment, Hemel Hempstead: Harvester Wheatsheaf.

Roos, S., Sandin, G., Zamani, B. and Peters, G., 2015. Environmental assessment of Swedish fashion consumption. Fivegarments—sustainable futures. Stockholm, Sweden: Mistra Future Fashion.

Shirvanimoghaddam, K., Motamed, B., Ramakrishna, S. and M. Naebe (2020) Death by waste: Fashion and textile circular economy case. Science of The Total Environment, 20 May, 718, 137317.

An Exploratory Study on the Role of Municipalities in the Circulation of Second-Hand Textile Products

9

Understanding the Process, Motives, and Challenges

MUTEBER TUZCU AND ÖZGE KIREZLI

Introduction

The circular economy is a relatively deep-rooted concept. However, it gained particular importance after the 1990s. The circular economy is a concept that includes planning, resourcing, procurement, production, and reprocessing creation in order to minimize waste and maximize both ecosystem functioning and human well-being (Murray, Skene and Haynes, 2015). Due to the increase in environmental concerns, various stakeholders like governments, businesses, and the academic environment are paying attention to the concept. According to the United Nations Environment Programme (UNEP) (2006), the circular economy helps to maintain the balance between economic development and the production of environmental sources. Geissdoerfer, Savaget, Bocken, and Hultink (2017) define the circular economy as a regenerative system in which minimizing resource input and waste, emission, and energy leakage come out by decelerating, closing, and narrowing materials and energy loops. Ellen MacArthur Foundation (Ellen MacArthur Foundation Circular Economy Glossary, 2001) defines the circular economy as

> A systems solution framework that tackles global challenges like climate change, biodiversity loss, waste, and pollution. It is based on three principles, driven by design: eliminate waste and pollution, circulate products and materials (at their highest value), and regenerate nature.

There are plenty of studies in order to define circular economy concepts (Peters, Weber, Guan and Hubacek, 2007; Geng and Doberstein, 2008; Yang and Feng, 2008; Gregson, Crang, Fuller and Holmes, 2015;

Stahel, 2016; Kirchherr, Reike, and Hekkert, 2017, Geissdoerfer et al., 2017, and Prieto-Sandoval, Jaca and Ormazabal, 2018). All of these authors define the circular economy on a close-loop system basis and mention the aim of the concept as minimizing waste and costs.

The circular economy creates a potential in three dimensions of sustainable development in regard to economic, environmental, and social (Korhonen, Honkasolo and Seppälä, 2018). Environmental wins, economical wins, and social wins occur at the same time in this system. For example, decreasing raw material and energy input affects diminishing raw materials and energy costs in economic regard. If the system works properly, the social benefits of the system would start by causing new employment opportunities and raising a sense of community and cooperation. During every step of the system, the value of materials is preserved and used more than one time.

Three principles, which are reduce, reuse, and recycle—also known as the three Rs—dominate the circular economy. Reduction implies a decrease in resources and energy consumption through the improvement of production efficiency. Reuse refers usage of by-products and waste from one phase of the production to the next phase. Lastly, recycling claims that output materials in any production process can be used in another process as input material (Zhu, Geng and Lai, 2010; Geisendorf and Pietrulla, 2017, and Heshmati, 2017).

In that study, we focused on local authorities' practices of the "reusability" dimension of the circular economy. Inspired by the recent news about heightened attention to collect and effectively distribute second-hand textile products, three research questions are formulated:

RQ1: What is the overall process the municipalities perform in the re-use of second-hand textile products? Who/What are the actors involved in that process?

RQ2: What are the motives of municipalities for supporting the circularity of second-hand textiles?

RQ3: What are the challenges of this process?

Literature Review

Post-Consumer Textile Products: Are They Really Waste or What?

One of the most polluting industries is the textile and fashion industry because of excessive amounts of water, chemicals, and energy usage (Riba,

Cantero, Canals and Puig, 2020). Not only, the rapid increase in population, but also the market's interest in fast fashion is triggering dramatic expanse in the textile industry. Especially brands like Zara, H&M, and Topshop encourage consumers' never-ending desire for buying items. Their seasonal collections have turned into monthly and weekly collections. More production in the industry leads to high consumption in the fashion industry, stimulating a vicious circle. For these reasons, the manufacturing of textiles pollutes nature irrevocably. Besides that, the textile industry has two facets of waste—pre-consumer waste and post-consumer waste. Thus, pollution caused by these phases contribute to each other.

Every day tons of apparel and textile waste are sent to landfills. This waste is the result of both pre-consumer and post-consumer waste. According to Büyükaslan, Jevsnik, and Kalaoğlu (2015), approximately half of textile and apparel waste constitute post-consumer disposal in Turkey. However, these materials have the potential to be recycled and reused. For example, post-consumer disposed textile waste is 100 percent recyclable. (Stall-Meadows and Peak, 2010). In linear economy systems, the value cannot be used again. In contrast, the circular economy highlights the importance of reusing and recycling as of creating value. The reuse of post-consumer textile waste is beneficial for social, environmental, and economic regards. Unfortunately, for instance, young female consumers in the United Kingdom are not aware of reuse and recycling due to a general lack of knowledge (Morgan and Birtwistle, 2009).

In order to capture value inside post-consumer apparel waste, Hawley (2006) has prepared a pyramid model for textile recycling categories. The pyramid has five layers that are in hierarchic order, based on the intensity of total post-consumer textile waste. These layers are used clothing markets, conversion to new products, wiping and polishing cloths, landfill and incineration, and diamonds. Used clothing markets that are the first step of the pyramid have abounding entities. In this category, nearly half of post-consumer waste is exported, sent to relief markets (second-hand markets in undeveloped countries), or sold in second-hand markets. The second layer is the conversion to a new product that benefits both the environment and the economy. At this stage, waste turns into new material preventing value loss for such waste through mechanical and chemical processes. In the third layer, there come from the second layer unusable cotton textile products like T-shirts are used as wiping and polishing cloth for industrial usage. In the following layer, the remaining waste from the bottom layers are sent to landfills and incineration for energy. The last step, which is the most profitable, and the least amount of textile waste is diamonds. In this category, there are haute couture appeals. Mostly designers, high-income level people, and vintage shop owners try to find and buy such appeals (Hawley, 2006).

Municipalities' Role in Circulation of Second-Hand Textile Products: Examining Practices

Countries such as China, Taiwan, Denmark, and the United Kingdom have strategic road maps highlighted by the government (Yuan and Moriguichi, 2008; Morgan and Bitwiste, 2009; Huang, Azevedo, Lin, Cheng and Lin; 2021, and Christensen, 2021). In order to apply these central strategic plans, municipalities become one of the significant parties in the system. There are three main reasons why municipalities have a significant role in the circular economy (Christensen, 2021). Firstly, the waste management systems are mostly under the responsibility of the local authorities. Before the importance of the circular economy was realized, municipalities had already been dealing with excessive amounts of waste. These high amounts of waste also pollute the environment. With the adoption of the circular economy approach, municipalities either re-distribute them or recycle large amounts of waste into useable materials. Thus, the damage to the environment is reduced. Secondly, circular economy practices fit environmental sustainability plans such as reduced air and water pollution and reduce energy usage. Thirdly, municipalities can act as a bridge between citizens and the private sector. The downside of the collection and managing is needed a high amount cost because of the heterogeneity character of municipalities' waste (Hoornweg and Bhada, 2012, and Zacho, Mosgard and Riisgaard, 2018).

In their research focusing on Danish municipalities' practices, Zacho and others (2018) delineated the reuse and the recycle processes of textile products. After the collection period, four streams of materials occur: items for reuse, cardboard, wood, and plastics. For items for reuse, several steps named sorting, testing, cleaning, price taking are applied and are sold. Cardboard and wood waste turn into raw materials—paper mill cardboard production and chipboard production through sorting, transporting, and shredding. These converting products are ready to reuse. For plastic waste, the process is a little bit complicated with respect to previous materials. Plastic waste is also sorted, transported, and shredded. Following the same processes as others, they are still counted as waste. Such waste should go through unknown processes in order to turn them into raw materials. The raw materials are again sorted, transported, and shredded for reuse. The first step—the collection step—is done by the largest recycling centers in a company. The remaining steps are performed at recycling facilities (Zacho et al., 2018).

Huang and others (2021) have investigated Taiwan's textile innovation strategies. They found six challenges—governmental and regulatory issues; economic and financial issues; technical issues; societal issues; infrastructural, supply chain, and market issues; and organizational and managerial issues—based on the previous literature. According to these authors, the infrastructural issue is the essential issue among these six issues because of the lack of a successful business model in this sector and information exchange among involved parties, and the outdated supply and demand networks. Among these six issues, societal issues are another important barrier to Turkish concept. The issue became a topic because of a lack of confidential public information, lack of social awareness, difficult to find personnel, and low customer demand for second-hand textile products.

Another perspective, regarding the role of local authorities, is the cooperation with non-governmental organizations (NGOs). Not only municipalities and government but also NGOs are also other key actors in the circular economy. As Hawley (2015) mentions, most developed countries export such idle products to undeveloped countries and gain economic and environmental benefits. On the other side, low purchasing power is one of the most fundamental issues in undeveloped countries. Thus, people in such countries can buy second-hand textile products less than new clothes (Brooks and Simon; 2012). The situation carries with both social and economic benefits to such people in undeveloped countries.

One of the most popular practices is from the United Kingdom, whereby donators can dispose of their own idle textile product to Oxfam, Salvation Army, and Cancer Research where these donated textile goods are sorted and sold in their stores. Useable textile products are also sold to Africa, East Europe, and Asia. These two sale systems are reuse practices. For reuse practice, the nonusable textile waste can be directly sent to landfills (Waste Online, 2008). In that sense, non-governmental organizations might be more active in some countries by not only disseminating these idle textile products to disadvantaged families but also selling and exporting to create economic benefits, as well.

Research Methodology

Research Design and Sampling

This study is designed as an exploratory qualitative study aiming to delineate the role of local authorities in the reuse and/or recycling processes of used textile products in terms of responsibilities, activities, cooperation with involving actors, motives, and challenges. To gain an insight into different procedures (clothes piggybanks, second-hand stores owned by municipalities, direct lending of the textile products to the needy people, etc.) on the issue, interviewees are purposively selected to cover diverse levels of local authorities including neighbourhood representatives, municipalities' social services, and metropolitan municipalities' social services. Judgemental and snowball sampling techniques (Flick, 2009; and Machado, Almeida, and Bragagnolo, 2019) are adopted to reach the target interviewees. The researchers sent invitation emails to municipalities' information, municipalities' social services, and municipalities' environmental department units for appointments both social services and units confirmed the appointment with a return rate of eight out of eleven (72 percent). The sample consists of seven participants (neighbourhood representative, three social aid specialists, a manager of the environmental production and control of the municipality, a head of the social aid department, and an environmental engineer of the municipality).

Data Collection Method, Data Collection Instrument, and Analysis

The interviews were performed in hybrid form due to pandemic conditions between August 25, 2021, and November 2, 2021. A total of seven semi-structured interviews, lasting between thirteen and forty-five minutes, were carried out face to face (five interviews) or via videoconferencing (two interviews). The authors developed a guide by reviewing the relevant literature and including self-developed questions, to maximize the information gathering process. However, on the basis of the flow of conversation, a semi-structured interview is adopted.

The interviews were recorded through audio devices (two videoconferencing, and five audios), enabling the transcription of the data. Data analysis followed a three-step process. First, during the interviews, the researchers take the accompanying notes separately. In the second step, after listening and transcribing the interviews, cross-examination is done to enrich the depth of the findings. Finally, a thematic analysis is performed to identify the emerging themes and key points.

Findings

Process of Re-distribution of Second-Hand Textiles

The overall circularity of these second-hand textile products is handled in diverse ways. However, covering all the phases, we believe the most

*Municipalities that are receiving the applications to use that system, control the appropriateness of these applicants by field trips. If the user is deemed as an appropriate one, an account is opened and items are delivered on a periodic basis.

Figure 9.1. Process of Re-distribution of Second-Hand Textiles
Source: Authors preparation

comprehensive process can be envisioned via Figure 9.1. Four main steps exist to delineate the whole process: collection, cleaning and sorting, displaying, and re-distributing based on the interviews. Basic explanations are provided in the next subsections.

STEP 1: COLLECTION. Collection of idle second-hand textile products is possible in three main streams. Donators might bring the items to the municipalities' facilities or call them to be collected directly from their homes by local authorities or deliver them to piggybanks by themselves.

Direct receiving of the items by the donators is quite straightforward. However, the piggybank channel requires one-step further examination. The first demand to donate their idle products come from the citizens to municipalities. Due to the limited budget of municipalities, they go bidding for firms that are willing to be part of these processes. During the bargaining period between firms and municipalities, the firm presented the profitable contract win the bidding and put their piggybanks based on municipalities' decisions. The idle textile products from piggybanks are collected from such firms. According to these contracts, firms might give back some portion of textile products to municipalities. Firms can also donate wheelchairs to the municipalities.

STEP 2: CLEANING AND SORTING. These collected second-hand textile products go to the second stage—warehouses of municipalities or warehouses—in three ways. Two of these three ways are performed by municipalities and the remaining are conducted by intermediaries. Even though two ways are applied by municipalities, the dominant party in the step is intermediaries. During this stage, small modifications such as buttonholing and unstitching on textile products are made on idle textile products. On the other hand, textile products that are not being able to repair are sent to landfills directly in order to recycle them.

STEP 3: DISPLAYING. Other textile products that are being cleaned and refurbished are set to meet new owners through municipalities' social stores. These stores are managed by municipalities whereby items are organized in a rigorous way. In that stage, some items might be refurbished to be lent to the users. In most municipalities, new coming textile products enter the store system manually.

STEP 4: RE-DISTRIBUTING. Disadvantaged people can visit these social stores and receive the needed textile products. Municipalities also render textile products up to houses of elderly and disabled people. Municipalities keep records of disadvantaged people's names, family members, addresses, and monthly income. These records come from both the e-government system and/or neighborhood representatives. Some municipalities in this research also can visit disadvantaged people's houses in order to identify what really needed is. Practices like visiting houses might prevent abuse of these textile products. According to the data and the practice, municipalities keep their own records in order to pursue visitors coming to their social market. Municipalities' records show the frequency of every single visitor to the social markets and units of products taken by disadvantaged people.

> The needy people cannot come to our store whenever they want and get as much product as they want. They can get a limited number of products based on income level and the number of people in their households. When they visit our store and obtain textile products, we inform them about the date when they have the right to get new products. (S3 and S6)

Motives

The motives in that circular system need to be analyzed in two dimensions; the nature of (pro-active or reactive nature of lending or receiving items) and the types of motives. First, it is quite evident that the primitive urges for social solidarity stem from the citizens. All of the interviewees stated that proactive donators reached them as of lending their usable but idle second-hand products. In return, local authorities responded to these attempts by enabling coordination.

> We act as a liaison between donators and users. With that aim, we utilize our resources either by directing human resources or designing operational processes. No matter what the item is, either it is a textile product, home appliances or book, if we can ensure the continuity of the system, all parties will be better off. (S6)

Secondly, as stated in the circular economy's motives, multiple economic, social, and environmental motives exist. However, our findings only validated the strong emphasis on "social motives," whereby both donators and local authorities are aware of how this system is contributing to disseminating a sense of collectivist tendencies in the society. Therefore, in that term, the direct motive is undoubtedly the social one. However, it is believed that the indirect effect of economic and environmental motives exists even though they are not overtly mentioned.

Challenges Emerging from the System

The third theme of the study focused on the challenges within the system. In the following parts, major challenges are explained. The most complicated phase of the whole process is undeniably the collection part because of the fact that idle wastes have heterogeneous characters. Multiple channels of collecting the items pose a challenge not only in coordination but also in control issues, as well. Three main challenges are identified in this step: lack of inspection, abuse of the items, and resource constraints.

LACK OF INSPECTION. Lack of inspection emerged as a challenge in two perspectives. The first one covers the lack of inspection in the items' perspective, whereby inappropriate items might enter the system, which increases the workload in the second phase of cleaning and sorting, as well. All the interviewees stated that, either in direct lending or municipality-aided deliveries, when the donators personally deliver the items, benefits overweigh the challenges. It is evident that one-to-one interaction in the delivery process increases the donators' sensitivity by perceiving the items as a form of self-expression. One interview stated that:

> When they personally bring the packages, donators really care not only about the usability of the items but also how they bring them. For example, you can really smell the detergent and see the color, these show how much concern donators put on the issue. (S1 and S3)

Unfortunately, the situation is severe in piggybanks. Due to a lack of inspection and automated like the process of giving away the items drops the quality of the items and under which conditions it is delivered. The local authorities emphasized that intermediaries faced with a large variety of items rather than second-hand textiles, including organic waste, books in bad condition due to contamination, and even electronics.

> I personally make spontaneous visits to intermediaries' warehouses and am stunned by the items piggybanks are filled with. Not to mention that citizens might be a bit confused to use these piggybanks as garbage, which increases the health risk intermediaries face with. Also, batteries and electronic products are thrown away. This is a problem. (S6)

The second dimension regarding the lack of inspection is about the balance of power between intermediaries and municipalities. Interviewees stated that the contract they have with the intermediaries is so restrictive that they are not fully aware of the full process and mostly renew the contract. The municipalities need to be empowered against intermediaries in terms of controlling and improving the value of the total circulation.

> Actually, we do not know on what basis these agreements were made. The contract was signed by the previous management team of the municipality. Due to the circumstances of the contract, we have to maintain this agreement even though the management team has changed. (S2)

SYSTEM ABUSE. System abuse is experienced in two aspects; unauthorized piggybank placements and free riders. As given in the process part, municipalities go out to bidding and clarify the conditions of how many piggybanks will be placed in which locations and when they will be collected, etc. However, some interviewees mentioned that they are facing unfair practices during field visits such as illegal piggybanks using municipalities or non-governmental organizations' names. For that reason, municipalities engage in routine control of the piggybanks and collect the unauthorized piggybanks with their enforcement power. In addition, the threat of free riders is a mentioned issue, whereby people invade piggybanks and collect the items by themselves, especially children. Either for selling or using these items, it is quite problematic to engage in such a physical risk for these free-riders. Also, the effort to systemize the circulation of these items, by prioritizing and validating the needy families, is damaged by these free riders. Even, the donators might be unwilling to contribute to the system by facing these abuses. Not all municipalities have the luxury to open and sustain a social store, which erodes to municipalities' responsibility as acting as a liaison between donators and users.

> We have observed that people actually abuse the products we distribute. For example, the products we distributed were sold in a bazaar within the borders of another municipality. People pretended to be in need of help and get products from our social store to sell. (S6)

RESOURCE CONSTRAINTS. The municipalities are getting help from intermediaries for the piggybank system. On the other hand, direct delivery from home imposes an extra burden on the local authorities. All of the interviewees expressed how in the collection of these items, they suffered from assigning an adequate number of human resources and vehicles.

> If only I had more resources, I will be organizing the process in a better way from collecting the items, refurbishing them and lending the items to the needy families. (S1)

> We have several responsibilities besides this one, and we need to use all of our resources way in the most efficient way. We lack the funds, personnel, and vehicles to focus periodic collection of these textile products, furniture and home appliances (S3)

> Besides our own personnel, we are getting help from volunteers and non-governmental organizations which increases the impact. As you know, municipal budget is limited; we try to maximize the social impact in every possible way via these voluntary efforts (S6)

Municipalities are quite passive in that cleaning and sorting phase due to the context of the contracts. Actually, it is stated that sorting and cleaning textile products are costly, and certainly more human resources are needed. In that perspective, involving intermediaries add value to the system. However, we believe in the whole process even though it might be stressful to manage that process, both the economic benefits and quality and quantity of the items might be enhanced if more control is ensured. Besides the benefits of contracts with firms, textile products in the diamond category might be lost or underestimated during cleaning and sorting. Moreover, in this general process, textile products in the diamonds category are not mentioned. As it is mentioned earlier, their profit margin is high. Municipalities can give importance to finding and separating these diamonds from ordinary idles.

In addition, unfluctuating demand in the social market might cause an overcrowded and inefficient flow of products. Limited numbers of employees in these stores cannot meet the demand. In addition to this, technology is not used efficiently in municipalities' stores. Every single entry is done manually in most cases. However, the usage of technology leads to least employees in these stores and helps to meet the high demand situation.

Moreover, finding the real "needy" people is a compelling job in this process. Unfortunately, some people request textile products, even though there are no real needy conditions. Such people abuse the system that is maintained by municipalities. So, identifying real disadvantaged people forces to the municipalities. In addition, recording information about needy people is time-wasting time and redundant paperwork. Thereafter, municipalities can benefit from such diamonds through sell them at the garage sales activity. The sale practices also provide extra income to municipalities. Furthermore, municipalities might face unwanted situations after so-called needy people get the products. Such people try to sell them.

Discussion and Conclusion

This study is designed to delineate the role of municipalities as of local authorities in the reusability of second-hand textile products. For that aim, the overall procedure of how these second-hand textile products are processed by the local authorities, their motives, and challenges in the system are explored.

Local authorities are fully aware of their role as social liaisons between disadvantaged people and donators. However, intermediaries are detected as extremely active involving actors. Not only the local authorities but also these intermediaries recognized the economic and social value of these idle textile products that triggered competition. In that perspective, it is believed that local authorities need to be more assertive to determine the contractual context for the piggybank system in terms of increasing transparency of the processes. Municipalities mostly renew the previous contract and fail to change the terms of the contract, which remains as insufficient to meet with the escalating needs of the people. Earlier, there were few firms and many municipalities, but now more intermediaries emerged, which triggered the competition.

Yet, we believe the role of metropolitan municipalities is underestimated. Due to the proliferation of the motives (intensifying social and economic impact and minimizing environmental impact), the activities performed by intermediaries might be undertaken by the metropolitan municipalities. Not only, it would guarantee the reach of operations (collection, cleaning, and sorting), the potential in the qualitative and quantitative nature of the idle textile products would be recognized.

Even though, convenience is a critical point in that process either collecting or re-distributing the items, lack of control especially in the collection of these items via piggybanks are quite challenging. Instead of increasing the prevalence of piggybanks, the social store model offers more advantages for the easy inspection of the items, avoiding contamination and offering a healthier environment whereby the users and municipality employees interact with each other. Since we detected a gap in better coordination of these idle items. If the metropolitan municipalities take in charge of this piggybank system and disable intermediaries, it would come with at least three anticipated benefits; re-directing to recycling facilities, garage sales of the "diamond category" and refurbishing the flawed items. Collection and sorting steps need to be handled by the metropolitan municipality, meaning that a better grasp of the process is ensured. In line with Christen's (2021) view on how municipalities need to be more active in both waste and usable material circularity, centralized sorting of the items will make it easier to detect, the usable and recyclable materials, which would establish a direct channel to sending items to recycling facilities. In most scenarios in this research, unusable textile waste is thrown away; metropolitan municipalities need to be more active in landfill operations.

In addition, another emerging theme is about the utilization of technology. As mentioned in Huang and others's (2021) study, infrastructural problems exert a challenge in the system, indicating insufficient usage of technology. In the first era of admitting the items via barcodes, will increase the municipalities' control over the overall cycle. If more technology would be used in social stores, in the displaying step, fewer employees will work there. This will cause a decrease in budgets and these employees can work other steps in these processes like cleaning and sorting products etc.

Like other studies, the study has several limitations. Depending on the interviews during the study, the main concern to apply re-use of textile products starts with social motives. However, the practices include more than social concerns such as economic and environmental concerns. In order to understand deeply these motives, the upcoming studies should investigate environmental focus on reduction, re-cycle, and re-use practices with different wastes such as yellow grease and waste batteries, and so on. The textile industry also harms the environment through the pre-consumer phase. The study examined only textile waste from the post-consumer phase. The industry has one of the most damaging outputs through manufacturing processes. Wastes re-cycling and re-usage from these processes should be visualized as well. Secondly, the number of interviewers in the study is quite limited. Also, other stakeholders, such as intermediaries, NGOs, donators, and users should be interviewed to gather a comprehensive evaluation of the system has more than one parties.

References

Brooks, A., and Simon, D. (2012). Unravelling the relationships between used-clothing imports and the decline of African Clothing Industries. Development and Change, *43*(6), 1265–90. https://doi.org/10.1111/j.1467-7660.2012.01797.x.

Büyükaslan, E., Jevšnik, S., and Kalaoglu, F.(2015). A sustainable approach to collect post-consumer textile waste in developing countries. Marmara University Journal of Science, *27*(3). https://doi.org/10.7240/mufbed.43773.

Christensen, T. B. (2021). Towards a circular economy in cities: Exploring local modes of governance in the transition towards a circular economy in construction and Textile Recycling. Journal of Cleaner Production, *305*, 127058. doi:10.1016/j.jclepro.2021.127058.

The Circular Economy Glossary. How to build a circular economy. (n.d.). Retrieved October 25, 2021, from https://ellenmacarthurfoundation.org/topics/circular-economy-introduction/glossary.

Environment, U. (n.d.). UNEP 2006 annual report. Retrieved October 10, 2021, from https://www.unep.org/resources/annual-report/unep-2006-annual-report.

Flick, U. (2009). An introduction to qualitative research (4th ed.). Sage Publications Ltd.

Geisendorf, S., and Pietrulla, F. (2017). The circular economy and circular economic concepts-a literature analysis and Redefinition. Thunderbird International Business Review, *60*(5), 771–82. https://doi.org/10.1002/tie.21924.

Geissdoerfer, M., Savaget, P., Bocken, N. M. P., and Hultink, E. J. (2017). The circular economy—a new sustainability paradigm? Journal of Cleaner Production, *143*, 757–68. https://doi.org/10.1016/j.jclepro.2016.12.048.

Geng, Y., and Doberstein, B. (2008). Developing the circular economy in China: Challenges and opportunities for achieving 'leapfrog development.' International Journal of Sustainable Development & World Ecology, *15*(3), 231–39. https://doi.org/10.3843/susdev.15.3:6.

Gregson, N., Crang, M., Fuller, S., and Holmes, H. (2015). Interrogating the Circular Economy: The Moral Economy of Resource Recovery in the EU. Economy and Society, *44*(2), 218–43. https://doi.org/10.1080/03085147.2015.1013353.

Hawley, J. M. (2006). Digging for diamonds: A conceptual framework for understanding reclaimed textile products. Clothing and Textiles Research Journal, *24*(3), 262–75. https://doi.org/10.1177/0887302x06294626.

Hawley, J. M. (2015). Apparel Recycling. Sustainable Apparel, 251–62. https://doi.org/10.1016/b978-1-78242-339-3.00011-x.

Heshmati, A. (2017). A review of the circular economy and its implementation. International Journal of Green Economics, 11(3/4), 251. doi:10.1504/ijge.2017.089856.

Hoornweg, D., Bhada, P., 2012. What a Waste. A Global Review of Solid Waste Management. Urban Development Series Knowledge Papers. http://dx.doi.org/10.1111/febs.13058.

Huang, Y.-F., Azevedo, S. G., Lin, T.-J., Cheng, C.-S., and Lin, C.-T. (2021). Exploring the decisive barriers to achieve circular economy: Strategies for the Textile Innovation in Taiwan. Sustainable Production and Consumption, *27*, 1406–23. https://doi.org/10.1016/j.spc.2021.03.007.

Kirchherr, J., Reike, D., and Hekkert, M. (2017). Conceptualizing the circular economy: An analysis of 114 definitions. Resources, Conservation and Recycling, *127*, 221–32. https://doi.org/10.1016/j.resconrec.2017.09.005.

Korhonen, J., Honkasalo, A., and Seppälä, J. (2018). Circular economy: The concept and its limitations. Ecological Economics, *143*, 37–46. https://doi.org/10.1016/j.ecolecon.2017.06.041.

Machado, M. A., Almeida, S. O., Bollick, L. C., and Bragagnolo, G. (2019). Second-hand fashion market: Consumer role in Circular Economy. Journal of Fashion Marketing and Management: An International Journal, *23*(3), 382–95. https://doi.org/10.1108/jfmm-07-2018-0099.

Morgan, L. R., and Birtwistle, G. (2009). An investigation of young fashion consumers' disposal habits. International Journal of Consumer Studies, *33*(2), 190–98. https://doi.org/10.1111/j.1470-6431.2009.00756.x.

Morseletto, P. (2020). Targets for a circular economy. Resources, Conservation and Recycling, 153, 104553. https://doi.org/10.1016/j.resconrec.2019.104553.

Murray, A., Skene, K., and Haynes, K. (2015). The circular economy: An interdisciplinary exploration of the concept and application in a global context. Journal of Business Ethics, *140*(3), 369–80. https://doi.org/10.1007/s10551-015-2693-2.

Peters, G. P., Weber, C. L., Guan, D., and Hubacek, K. (2007). China's growing co2 emissionsa race between increasing consumption and efficiency gains. Environmental Science & Technology, *41*(17), 5939–44. https://doi.org/10.1021/es070108f.

Prieto-Sandoval, V., Jaca, C., and Ormazabal, M. (2018). Towards a consensus on the circular economy. Journal of Cleaner Production, *179*, 605–15. https://doi.org/10.1016/j.jclepro.2017.12.224.

Riba, J.-R., Cantero, R., Canals, T., and Puig, R. (2020). Circular economy of post-consumer textile waste: Classification through infrared spectroscopy. *Journal of Cleaner Production*, *272*, 123011. https://doi.org/10.1016/j.jclepro.2020.123011.

Stahel, W. R. (2016). The circular economy. Nature, *531*(7595), 435–38. https://doi.org/10.1038/531435a.

Stall-Meadows, C., and Peek, G. (2010). Recycled household textiles and clothing. Oklahoma Cooperative Extension Service Fact Sheet T-4318.

Waste Online (2008) Waste Online. [WWW document]. URL http://www.wasteonline.org.uk/index.aspx (accessed on 9 November 2008).

Yang, S., and Feng, N. (2008). A case study of industrial symbiosis: Nanning Sugar Co., Ltd. in China. Resources, Conservation and Recycling, *52*(5), 813–20. https://doi.org/10.1016/j.resconrec.2007.11.008.

Yuan, Z., Bi, J., and Moriguichi, Y. (2008). The circular economy: A new development strategy in China. Journal of Industrial Ecology, *10*(1–2), 4–8. https://doi.org/10.1162/108819806775545321.

Zacho, K. O., Mosgaard, M., and Riisgaard, H. (2018). Capturing uncaptured values—a Danish case study on municipal preparation for reuse and recycling of waste. Resources, Conservation and Recycling, *136*, 297–305. https://doi.org/10.1016/j.resconrec.2018.04.031.

Zhu, Q., Geng, Y., and Lai, K.-H. (2010). Circular economy practices among Chinese manufacturers varying in environmental-oriented supply chain cooperation and the performance implications. Journal of Environmental Management, *91*(6), 1324–31. https://doi.org/10.1016/j.jenvman.2010.02.013.

Financial Aspects of Environmental, Social and Governance Criteria 10

Turkish Case

KUBILAY DÜZGIT AND MEHMET NURI ÖNORTAÇ

Introduction

In the recent years, company's environmental, social, and governance (ESG) criteria has become an important part of the decision-making process of any financial statement user. The ESG criteria is an indicator of a company's environmental and social responsibility (Friede, Busch, and Bassen, 2015). As global warming becomes a topic highly debated, and renewable energies getting financially viable. The public requires the corporations to be held responsible of their economic activities for the consequences to nature and humanity. In Turkey, the abundance of news about natural disasters, pollution, income inequality and increasing unemployment rates of university graduates are worrying. That's why the assessment of corporation's social and environmental responsibility is becoming an important aspect (Aktaş, Kayalidere, and Karğin, 2013).

With the launch of several ESG funds by commercial banks in Turkey, socially responsible investments and impact investments are becoming much more accessible to the public. With these types of funds rolling out and as a result, socially responsible investments and impact investments getting more common, companies need to prioritize their practices in ESG areas to increase their scores and lift their image in the eyes of the stakeholders (Kuzucu, 2018). There are many studies that examined the relationship between ESG criteria and corporate financial performance in developed countries. But in the case of Turkey, an insufficient number of studies have been conducted to reach a consensus about the relationship between ESG criteria and corporate financial performance of Turkish companies.

The purpose of this research is to close the literature gap concerning the relationship between ESG criteria and firm financial performance in Turkish companies. To reveal this relationship panel, data analysis method was used as a means to examine if a company's ESG criteria rankings had any effect on their financial performance. As the analysis sample, non-financial firms quoted on Borsa Istanbul Stock Exchange were chosen between the years 2011 and 2020, since the aim is to contribute to the literature in regards of the relationship in Turkey.

The structure of this study is as follows: the literature review section is devoted to the review of related literature and empirical findings regarding the relationship between environmental, social, and governance criteria and financial performance. In Section 3, the hypothesis has been developed in accordance with previous literature. Methodology and analysis are extensively described step by step in Section 4. Results are interpreted in Section 5. Finally, in Section 6, concluding remarks have been mentioned.

Literature Review

In literature there have been many studies carried out, both empirical and theoretical, to try to understand the relationship between ESG criteria and corporate financial performance. Many of these studies examined the potency, sequence, and relationship causality, which resulted in both supportive and conflicting evidence (Yoo and Managi, 2021). Thus, it is difficult to give a final verdict to the relationship between the variables beforehand, as empirical studies on the relationship between ESG criteria and corporate financial performance have not shown definite results. There are three main categories of findings in the literature. First being a positive causal relationship, second being a negative causal relationship, and third being a neutral relationship. In the following paragraphs all three types of related evidence are reviewed.

Based on our literature review, some of the empirical research argue about the positive link between ESG criteria and financial performance (Landi and Sciarelli, 2019; Zhao, et al., 2018; Dalal and Thaker, 2019). This positive link has been explained by the hypothesis of social impact, which was based on the instrumental stakeholder theory (Donaldson and Preston, 1995). According to these theories, addressing the needs of various stakeholder parties would influence financial results to the better based on high productivity and efficiency. In the other hand, ignoring the demands of stakeholders may harm company's corporate financial performance. Firms that are addressing these environmental, social and governance criteria will also boost their public image that will protect the businesses from unexpected threats and lay the groundwork for new opportunities.

Velte's (2017) study supports these theories. Companies listed on DAX30, TecDAX and MDAX between 2010 and 2014 were examined to reveal the relationship between ESG criteria and financial performance. ESG scores were gathered from the Thomson Reuters database. Return on assets (ROA) and Tobin's Q were chosen as the representation of financial performance. Regression and correlation analysis' were carried out in order to investigate this relationship. The findings of this study suggests that ESG criteria has a positive impact financial performance. And compared to other categories of ESG governance had the strongest link between financial performance.

As for the negative ESG criteria and financial performance link, various studies have disclosed evidence (Balatbat, Siew, and Carmichael, 2012; Duque-Grisales and Aguilera-Caracuel, 2021; Saygili, Arslan, and Birkan, 2021). They argued that factors that increase ESG scores are not cost effective and that they increase the workload on administration. This is backed by the trade-off hypothesis, where it is proposed that the more a company engages with activities that result on an increase on ESG scores, the worse their corporate financial position is going to be (Myers, 1984). They argue that the sole purpose of a company is to maximize shareholder's wealth. Because these practices do not contribute to the companies' economic activities. It is seen as a waste of the investment of the shareholders.

Folger-Laronde, Pashang, Feor, and ElAlfy (2022) argues that during the COVID-19 pandemic ETF's that received high ESG ratings also incurred higher losses compared to ETF's that received low ESG ratings. They employed ANOVA and multivariate regression models to assess the performance of the ETF's and the relationship of their ESG ratings with their financial performance. ESG ratings proved to have a negative causal relationship with financial performance. So in a market crash like that caused by the COVID-19 pandemic, it could be said that holding high ESG rated ETF's could be a risky practice.

The recent work of El Khoury, Nasrallah, and Alareeni (2021) touched on the ESG criteria and financial performance's neutral relationship. They investigated the effects of ESG criteria on banks performance in the Middle East, North Africa and Turkey (MENAT) region between the years of 2007 and 2019. The financial performance of a total of forty-six banks was reviewed in order to give a verdict. As proxies for financial performance return on assets (ROA) and return on equity (ROE) were used. The results of the analysis show that there is no definitive effects of ESG on financial performance. For the analysis period evidence shows that ESG and financial performance have a positive relationship in the lower levels of investments. But this effect gives place to a negative relationship when the investment levels increase.

Hypothesis Development

Numerous empirical studies researched the financial aspects and the effects of ESG criteria on a firm's corporate financial performance. The results of these studies are conflictive. Some of them had results that support a positive relationship, where some of the studies have supportive evidence of a negative or even neutral relationship.

In this chapter, we are going to hypothesize like many other researchers in the field (López-Toro, Sánchez-Teba, and Benítez-Márquez, 2021; Zubeltzu-Jaka, Andicoechea-Arondo, and Alvarez Etxeberria, 2018; Garcia and Orsato, 2020), that there is a positive relationship between environmental, social, and governance criteria and corporate financial performance. Below there are two hypotheses generated in accordance with the analysis models later described in the chapter:

Hypothesis 1: There is a positive relationship between ESG scores and Return on Assets

Hypothesis 2: There is a positive relationship between ESG scores and Return on Equity

Methodology

The dataset for this research consists of fifteen non-financial Turkish companies for the period of 2011 to 2020. The data used in our analysis was obtained from Thomson Reuters Database. STATA 16 statistics software was used to run the panel data analysis. The data is strongly balanced (there is no missing data across all the 150 firm-year observations). The study aims to investigate the relationship between ESG criteria and corporate financial performance. Two models were used to test this connection. One where the dependent variable was measured using return on assets, and one where it was measured by return on equity (Siew, Balatbat, and Carmichael, 2013). Both ratios are a common profitability measurement and comparison tool in finance. Environmental, social and governance criteria is proxied by ESG scores (Aybars, Ataünal, and Gürbüz, 2019), which are calculated and published by Thomson Reuters. Other factors, namely, firm size and leverage, which can have an impact on a firm's financial

Table 10.1. The Variables and Their Abbreviations Used in This Study

Variable	*Abbreviation*	*Definition*
	Dependent Variables	
Return on Assets	ROA_{it}	Net income/total assets for firm i at year t
Return on Equity	ROE_{it}	Net income/shareholder's equity for firm i at year t
	Independent Variable	
Environmental Social and Governance Criteria	ESG_{it}	ESG score for firm i at year t
	Control Variables	
Firm Size	$SIZE_{it}$	Natural logarithm of total assets for firm i at year t
Leverage	$LEVERAGE_{it}$	Total debt/Total assets for firm i at year t

Source: Authors preparation

performance are included in our analysis in the form of control variables. Variables used in this study with their abbreviations and definitions are summarized in Table 10.1 above.

Our research hypotheses were tested to find out the relationship between environmental, social and governance criteria and financial performance. Models below shows the functional representation of our hypotheses:

Model 1:

$$ROA_{it} = \beta_0 + \beta_1 ESG_{it} + \beta_2 SIZE_{it} + \beta_3 LEVERAGE_{it} + u_{it}$$

Model 2:

$$ROE_{it} = \beta_0 + \beta_1 ESG_{it} + \beta_2 SIZE_{it} + \beta_3 LEVERAGE_{it} + u_{it}$$

In our analysis, we looked for multicollinearity, which can be described as correlation between independent and control variables. Multicollinearity, if present in the model, can severely distort the analysis results. In our case we tested our model using Variance Increasing Factor (VIF) and the results shows that there is no multicollinearity in the model. Secondly Likelihood-ratio test was run to identify the estimator to be used. Results indicated that for both models, there was a presence of unit effects. For our first model, Hausman test demonstrated that the unit effects were at random, and for the second model the results show that the unit effects were fixed.

To test the deviations from the basic assumptions of panel data analysis, firstly we ran a Jarque-Bera normality test. According to the test we concluded that the dataset was normally distributed. For heteroskedasticity we used Wald test. For both models results indicated the presence of heteroskedasticity. Pesaran, Friedman and Frees tests were used to look for cross-sectional independence, and it was revealed that the dataset was free from cross-sectional independence. Lastly, the test for autocorrelation of Baltagi and Durbin-Watson were performed. According to the test results both models suffered from autocorrelation. Under the deviations of heteroskedasticity and autocorrelation, a robust standard errors estimator (Tatoğlu, 2020), Driscool-Kraay was used for the models to conduct the regression analysis.

Results

Table 10.2 portrays the results of our first model. The findings support our initial hypothesis, which stated the positive relationship of environmental, social and governance criteria and corporate financial performance. The results demonstrate the positive and significant impact of ESG scores on financial performance of selected non-financial Turkish companies quoted on Borsa Istanbul Stock Exchange for years 2011 to 2020.

The control variables, both firm size and leverage had statistically significant impact on corporate financial performance. Firm size had a negative influence on financial performance, where leverage had a positive influence for the observation period of the selected firms.

Below, Table 10.3 displays the results of our second model. In line with our first model the results of the second model also shows the positive

Table 10.2. Results of The Random-Effects Panel Data Analysis for The Dependent Variable ROA

Regression with Driscoll-Kraay standard errors			
Dependent Variable: ROA			
Variables	*Coefficients*	*Standard Errors*	*T Statistics*
ESG	0,0002408**	0.0000831	2.90
SIZE	–0,0355554***	0.0045096	–7.88
LEVERAGE	0,0107689*	0.0052418	2.05
Constant	0,4043691***	0.0366776	11.02
Number of observations	150		
Number of groups	15		
Wald chi2	86,93		
Prob > chi2	0,0000		

*Legend: *p < 0,1; **p < 0,05; ***p < 0,01*

Source: Authors calculations

Table 10.3. Results of The Fixed-Effects Panel Data Analysis for The Dependent Variable ROE

Regression with Driscoll-Kraay standard errors			
Dependent Variable: ROE			
Variables	*Coefficients*	*Standard Errors*	*T Statistics*
ESG	0,0019648**	0.0006008	3.27
SIZE	−0,0997099***	0.0210485	−4.74
LEVERAGE	−0,0117475	0.0198879	−0.59
Constant	1,139937***	0.2132787	5.34
Number of observations	150		
Number of groups	15		
F	9.78		
Prob > F	0,0034		

*Legend: *p < 0,1; **p < 0,05; ***p < 0,01*

Source: Authors calculations

relationship of environmental, social and governance criteria and corporate financial performance. The results again demonstrate, the positive and significant impact of ESG scores on financial performance of selected non-financial Turkish companies quoted on Borsa Istanbul Stock Exchange for years 2011 to 2020.

In our second model only one of our control variables, firm size, was found to have a statistically significant influence on financial performance. It had a negative effect on financial performance in line with our first model for the selected companies during the observation period.

The results of our research, which portrays the environmental, social and governance criteria as an impact factor of corporate financial performance for the selected companies quoted in Borsa Istanbul Stock Exchange is supportive with other studies applied in different regions and countries around the world (López-Arceiz, et al., 2018; Abdi, Li, and Càmara-Turull, 2021; Almeyda and Darmansya, 2019).

Concluding Remarks

As discussed extensively, environmental, social, and governance field has been a hot topic for both academicians and administrative bodies. Academicians have found various results such as positive and negative relationship between the ESG criteria and corporate financial performance. In this study we examined the effect of ESG criteria on financial performance of fifteen non-financial companies quoted on Borsa Istanbul Stock Exchange, and we hypothesized that the relationship between environmental, social and governance criteria and corporate financial performance was positive.

The empirical evidence of our research analysis is in line with the hypothesis generated, and it reveals the positive relationship, supportive with other works around the world. Hence, it could be concluded that environmental, social and governance criteria has a positive effect on corporate financial performance of the selected Turkish companies quoted on Borsa Istanbul Stock Exchange for the years 2011 to 2020.

Companies can take in to account these results in their future activities. As it is verified by our analysis, the ESG criteria has a positive impact on financial performance. These results can be interpreted from two different perspectives. It can be said that the investors and consumers in Turkey are becoming more selective and conscious about their investment and consuming behaviors. They are rewarding the companies that are environmentally and socially responsible and punishing the ones who are not. And for the companies it is beneficial, that these practices improve firm's financial performance and the reputation of the company. In this manner Turkish companies should focus on improving their environmental, social and governance scores to be long-lasting and competitive.

For further research, the factors that affect environmental, social and governance criteria can be examined to find out which factors have the most influence on these scores and to give weight to those factors to improve financial performance.

References

Abdi, Y., Li, X., and Càmara-Turull, X. (2021). Exploring the impact of sustainability (ESG) disclosure on firm value and financial performance (FP) in airline industry: The moderating role of size and age. *Environment, Development and Sustainability*, 1–21.

Aktaş, R., Kayalidere, K., and Karğin, M. (2013). Corporate sustainability reporting and analysis of sustainability reports in Turkey. *International Journal of Economics and Finance*, 113–25.

Almeyda, R., and Darmansya, A. (2019). The influence of environmental, social, and governance (ESG) disclosure on firm financial performance. *IPTEK Journal of Proceedings Series*, 278–90.

Aybars, A., Ataünal, L., and Gürbüz, A. O. (2019). ESG and financial performance: impact of environmental, social, and governance issues on corporate performance. *Handbook of Research on Managerial Thinking in Global Business Economics*, 520–36.

Balatbat, M., Siew, R., and Carmichael, D. (2012). ESG scores and its influence on firm performance: Australian evidence. *Australian School of Business School of Accounting, School of Accounting Seminar Series Semester*, Vol. 2, pp. 1–30.

Dalal, K. K., and Thaker, N. (2019). ESG and corporate financial performance: A panel study of Indian companies. *IUP Journal of Corporate Governance*, 44–59.

Donaldson, T., and Preston, L. E. (1995). The stakeholder theory of the corporation: Concepts, evidence, and implications. *Academy of management Review*, 65–91.

Duque-Grisales, E., and Aguilera-Caracuel, J. (2021). Environmental, social and governance (ESG) scores and financial performance of multilatinas: Moderating effects of geographic international diversification and financial slack. *Journal of Business Ethics*, 315–34.

El Khoury, R., Nasrallah, N., and Alareeni, B. (2021). ESG and financial performance of banks in the MENAT region: concavity–convexity patterns. *Journal of Sustainable Finance & Investment.*

Folger-Laronde, Z., Pashang, S., Feor, L., and ElAlfy, A. (2022). ESG ratings and financial performance of exchange-traded funds during the COVID-19 pandemic. *Journal of Sustainable Finance & Investment*, 490–96.

Friede, G., Busch, T., and Bassen, A. (2015). ESG and financial performance: aggregated evidence from more than 2000 empirical studies. *Journal of Sustainable Finance & Investment*, 210–33.

Garcia, A. S., and Orsato, R. J. (2020). Testing the institutional difference hypothesis: A study about environmental, social, governance, and financial performance. *Business Strategy and the Environment*, 3261–72.

Kuzucu, N. (2018). Socially responsible finance: A current literature survey on the concepts and the approaches. *PressAcademia Procedia*, 51–55.

Landi, G., and Sciarelli, M. (2019). Towards a more ethical market: the impact of ESG rating on corporate financial performance. *Social Responsibility Journal.*

López-Arceiz, F. J., Bellostas-Pérezgrueso, A. J., Moneva-Abadía, J. M., and Rivera-Torres, M. P. (2018). The role of corporate governance and transparency in the generation of financial performance in socially responsible companies. *Spanish Journal of Finance and Accounting/Revista Española de Financiación y Contabilidad*, 44–80.

López-Toro, A. A., Sánchez-Teba, E. M., and Benítez-Márquez, M. D. (2021). Influence of ESGC Indicators on Financial Performance of Listed Pharmaceutical Companies. *International Journal of Environmental Research and Public Health*, 45–56.

Myers, S. C. (1984). Capital structure puzzle. *NBER Working Paper.*

Saygili, E., Arslan, S., and Birkan, A. O. (2021). ESG practices and corporate financial performance: Evidence from Borsa Istanbul. *Borsa Istanbul Review.*

Siew, R. Y., Balatbat, M. C., and Carmichael, D. G. (2013). The relationship between sustainability practices and financial performance of construction companies. *Smart and Sustainable Built Environment.*

Tatoğlu, F. Y. (2020). *Panel Veri Ekonometrisi*. Istanbul: Beta Yayınları.

Velte, P. (2017). Does ESG performance have an impact on financial performance? Evidence from Germany. *Journal of Global Responsibility.*

Yoo, S., and Managi, S. (2021). Disclosure or action: Evaluating ESG behavior towards financial performance. *Finance Research Letters*, 102–8.

Zhao, C., Guo, Y., Yuan, J., Wu, M., Li, D., Zhou, Y., and Kang, J. (2018). ESG and corporate financial performance: Empirical evidence from China's listed power generation companies. *Sustainability*, 2607.

Zubeltzu-Jaka, E., Andicoechea-Arondo, L., and Alvarez Etxeberria, I. (2018). Corporate social responsibility and corporate governance and corporate financial performance: Bridging concepts for a more ethical business model. *Business Strategy & Development*, 214–22.

Quantifying Degree of Circularity in Turkey 11
Material Flow Analysis

TULIN JOHANSSON AND MERAL ÖZALP

Introduction

Shifts toward circular economy (CE) have gained momentum in recent years due to the accelerated increase in the use of global raw material and its subsequent environmental and economic consequences. It is obvious that the capacities of ecosystems are no longer sufficient to absorb the increasing outflows of the material and energy use of the industry (Haas et al., 2015). The logic underlying linear economy "take-make-consume-dispose" has been replaced with "produce-use-reuse," a key concept of CE (Geisendorf and Pietrulla, 2017; EASAC, 2016; Haas et al., 2015; Alwood et al., 2010). Many governments and international organizations promote CE as a tool to mitigate greenhouse gas emissions (Haas et al., 2015) by introducing new business models, recycling, resource efficiency, and new consumption patterns (European Commission, 2019).

The mainstream literature lacks a singular, unique definition to the notion of CE (Geissdoerfer et al. 2017; Geisendorf and Pietrulla, 2017; Bocken et al. 2016; Lieder and Rashid 2016). According to the United Nations (UN) GEO5 report regarding CE, designs of material flows can be distinguished in two manners; biological nutrients re-entering to biosphere and materials for reuse and recycle within the economy (Haas et al. 2015; GEO5, 2012). In CE the products and materials are reused, refurbished, repaired or recycled, and waste becomes a resource (Veral, 2018; Jurgilevich et al., 2016). One of the most recognized definitions of CE is Ellen MacArthur Foundation's (EMF) definition, which introduces CE as a *restorative* and *regenerative* industrial system with the use of renewable energy and removing the use of toxic chemicals where the aim is to design-out waste (EMF, 2015). In this perspective, the notions of "eco-design, repair, reuse, renewal, remanufacturing, product sharing, waste prevention, and waste recycling" become significant while material losses are reduced through minimization of energy recovery from non-re-cyclable waste, landfill, and incineration (Sapmaz Veral and Yigitbasioglu, 2019; European Commission, 2015; EMF, 2015). Another well-known definition of CE was made in the EU Action Plan that focused on maintaining the value of the product as long as possible in the economy by minimizing the waste of resources and creating further value of the product that reached to the end of its life (European Commission, 2015). Arruda and others (2021) have reviewed the most cited articles in scientific journals, which state that in literature CE is a condition for sustainability with the greatest probability of favorable economic development and concluded that to reach circularity, the useful life of the resources must be extended.

The origins of CE can be traced back to Boulding's book *The Economics of the Coming Spaceship Earth* (1966), which attracts attention to the necessity of circular systems to guarantee human life on earth for the future (Geisendorf and Pietrulla, 2017; Geissdoerfer et al., 2017). However, the introduction of the concept of CE is mostly attributed to Pearce and Turner (1990) (Geissdoerfer et al., 2017; Ghisselini et al., 2016, Su et al., 2013; Andersen, 2006) that supported the idea of circularity because of unsustainability of traditional linear system that excludes recycling (Geissdoerfer et al., 2017). The first step for the implementation of CE is most commonly known as the 1972 BGBI Waste disposal act policy in Germany (Geisendorf and Pietrulla, 2017). Germany also played a frontier role in including the CE in national laws with the "Closed Substance Cycle and Waste Management Act" in 1996 (Su et al.2013; Geissdoerfer et al., 2017). Japan included the concept in 2002 with "The Basic Law for Establishing a Recycling-Based Society" (Geissdoerfer et al., 2017) and China in 2009 with "Circular Economy Promotion Law of the People's Republic of China" (Geissdoerfer et al., 2017; Lieder and Rashid, 2016). EU's CE strategy was announced in 2015 "Closing the loop—An EU Action Plan for the Circular Economy" (Geissdoerfer et al., 2017; European Commission, 2015).

The first CE action plan of the EU aimed to maintain the value of products, materials, and resources in the economy for as long as possible and minimize waste generation. This action plan was expected to increase job creation, stimulate economic growth and investment while also aiding to transition towards a carbon neutral, resource efficient and competitive economy (European Commission, 2015). Report of the commission on implementation of CE action plan (2019) assessed that the transition toward CE is accelerated due to the measures taken by the action plan. According to the report, the path towards accelerating transition toward the circularity is dependent on innovation and investments with a strong stakeholder engagement through circular design of products, empowering consumers (encouraging consumers to change their consumption patterns), turning waste into resources (efficient waste management systems), closed-loops for recovered materials (usage of secondary raw materials), and strategy for plastics (European Commission, 2019). EU's new CE Action Plan published in March of 2020 focused on the sectors that use majority of resources and with the high potential of circularity through the transition toward CE. The targets of CE Action Plan of EU (2020) are in line with Green Deal Targets and the new industrial strategy that aims the green economy-based policy changes (T.C. Ticaret Bakanlığı, 2021; European Commission, 2020).

In order for the assessment of CE strategies, indicators and monitoring tools are required to measure the degree of loop closing (Mayer et al. 2018; Krausmann et al. 2017; EASAC, 2016). The "Indicators of Circular Economy" report of the European Academies' Science Advisory Council (EASAC) collected currently used indicators to a CE from the resources UNEP, UNDP, Global Reporting Initiative (GRI), Yale and Columbia universities, World bank, OECD, Eurostat, EMF, and so on. The report classified the indicators related to CE under sustainable development, environmental, material flow, societal behavior, organizational behavior, and economy performance headings (EASAC, 2016). Economy-wide material flow accounts (EW-MFA) compiled by Eurostat appear to provide more specific information (EASAC, 2016), as data extracted under the MFA framework establishes a solid base for the discussion of policies for sustainable resource usage when assessed with traditional economic and demographic indicators (Fisher Kowalski et al., 2011).

There is a substantial body of empirical studies conducted utilizing EW-MFA framework. Mayer and others (2018) have measured the circularity of total material and waste flows and their socio economic and ecological loop closing through Sankey diagrams in European Union (EU28) for the year 2014. Bringezu and others (2015) investigated how a sustainability path for global material resources use can be formed based on existing EW-MFA indicators. Haas and others (2015) attempted to assess the key indicators of circularity based on the EW-MFA data extracted from Eurostat for the global economy and EU.

In practice, recycling still seems to be the key strategy to reach CE (Haas et al., 2015). Although recycling leads a significant reduction in virgin raw material use, where the recycling is very advanced (such as with metals, paper, glass and so on), it is also debated throughout the literature that recycling alone cannot establish the desired level of CE (Haas et al., 2015; Geisendorf and Pietrulla, 2017) due to the loss of quantity and quality created by the loop, which again creates the need for new materials and energy (Cullen, 2017). Beside closing loops through recycling and reuse, fossil energy sources must be replaced with renewable energy sources and overall resource consumption should be reduced (Haas et al., 2015; Geisendorf and Pietrulla, 2017).

CE policy of member countries and candidate countries of the EU, such as Turkey, are expected to be in line with EU CE plan. Chapter 27 regarding environmental negotiations between Turkey and EU makes EU's environment policies significantly important for Turkey (Sapmaz Veral and Yigitbasioglu, 2019). In the context of sustainability, the close follow-up and compliance with the regulations and targets set by the EU is crucial in terms of maintaining free trade between Turkey and the EU through harmonization with EU's technical legislation. Turkey's "Green Deal Action Plan" was published in July 2021, and aims to ensure the adaptation of the policies to fight against climate change and to maintain Turkey's competitiveness in exports (T.C. Ticaret Bakanlığı, 2021). Hence, it is essential to quantify and monitor the circularity of Turkey to best assess the effectiveness of current policies and identifying the possible future action plans to improve the circularity in the country.

This study aims to quantify the circularity of Turkey's economy over time at the macro-economic level by measuring the "circular material use rate (CMU)" developed by Eurostat (European Commission, 2018a). CMU measures the contribution of recycled materials to overall materials use, and thus it is identified as one of the relevant indicators of CE under the thematic area of "secondary raw materials" (European Commission, 2018). Furthermore, CMU has close connection to the Sankey diagram, which provides the visual presentation of material flows entering and existing the economy, the processing of materials into products, accumulation of stocks, becoming waste, and finally re-feeding into economy or discharging to the environment. Therefore, CMU is an appropriate indicator for closing the loop.

The remainder of this article is organized as follows. The next section briefly describes the data sources and the methodology used in deriving the CMU indicator. This is followed by presentation and interpretation of relevant data variables as well as the empirical results of the CMU indicator for Turkish economy. The final section draws upon some general conclusions along with the limitations of the study sketching out the way ahead.

Data Sources and Methodology

Data Sources

This chapter mainly utilized the Eurostat MFA [ENV_AC_MFA] (Eurostat, 2021b) and treatment of waste, by waste category [ENV_WASTRT] (Eurostat, 2021c) data sets that are extracted from Eurostat for the years 2014, 2016, and 2018, where the 2018 data set is the most recent data available for Turkey at the time of this writing. In addition, while the MFA are available for every year, the waste data is only available every second year. Hence, only even-numbered years are analyzed in this study. Furthermore, due to appeared inconsistency in the 2012 waste treatment data set, this study has not included the 2012 and previous years.

The MFA data set, which represents the input side of the MFA gives detailed information on Domestic Extraction, Imports, Exports, and Domestic Material Consumption for four different materials categories that are 1. Biomass, 2. Metal ores, 3. Non-metallic minerals, and 4. Fossil energy materials/carriers as well as total materials in each category (Eurostat, 2021a; European Commission, 2018b).

[ENV_WASTRT] data set is employed to analyze the output side of the Material Flow Analysis. Unfortunately, detailed waste data is not available for Turkey and many data items are either missing or specified as confidential. Hence, conducting detailed Material Flow Analysis is not possible for the case of Turkey. However, aggregated data for waste management operations are available for 2014, 2016, and 2018, and thus has been the base of this study together with MFA data set.

In addition, transboundary shipments of notified waste by partner, hazardousness, and waste management operations data set (Eurostat, 2021f) are used in estimating imported and exported waste bound for recovery in and out of Turkey.

Methodology

The methodology used in this study is completely inspired by the approach developed by Eurostat (European Commission, 2018a). CMU measures the shares of materials recovered and re-fed into the economy and it is calculated as a ratio of recycled materials (U) to processed materials (PM).

$$CMU = U/PM \quad (1)$$

PM are all primary and secondary materials used in the economic system and calculated as:

$$PM = \text{Domestic Consumption Materials } (DMC) + U\,, \tag{2}$$

DMC includes materials taken into use from nature in Turkey, to which imported goods is added and from which the exports are subtracted.

$$DMC = \text{Domestic Extraction } (DE) + \text{Imports–Exports}. \tag{3}$$

U are secondary materials where recycling is defined as 'any recovery operation by which waste materials are reprocessed into products, materials or substances whether for the original or other purposes and is calculated as

$$U = RCV\_R - IMPw + EXPw\,, \tag{4}$$

where *RCV-R* is recycled domestic waste excluding the recovery for energy, *IMPw* is amount of imported waste bound for recovery, and *EXPw* is the amount of exported waste bound for recovery.

CMU is the secondary materials as share of *PM* and can be expressed as:

TEXTBOX 11.1
Secondary materials

$$CMU = \frac{(RCV\_R - IMPw + EXPw)}{DMC + (RCV\_R - IMPw + EXPw)}$$

Source: The methodology used in this study is completely inspired by the approach developed by Eurostat (European Commission, 2018a).

Results and Interpretation

The analysis of ENV_AC_MFA data indicates that between 2014 and 2018, DE has increased between 874,122 and 908,534 thousand tons representing a 3.93 percent increase, while DMC has increased between 932,046 and 987,776 thousand tons representing a 5.97 percent increase between 2014 and 2018. Growing trends of DE is the growing pressure on the natural resources in the country. Figure 11.1 reveals that total DE increased by 5 percent but it decreased 1 percent in 2016 and 2018. Much of the decrease in DE is due to a 30 percent decrease in metal ores extraction in 2014 and 2016. However, biomass and fossil energy materials/carriers extraction has been increasing in the country. Biomass extraction has increased by 16 percent and 2 percent and fossil energy carriers increased by 13 percent and 15 percent in 2016 and 2018 respectively. Non-metallic minerals extraction has been steady over the years.

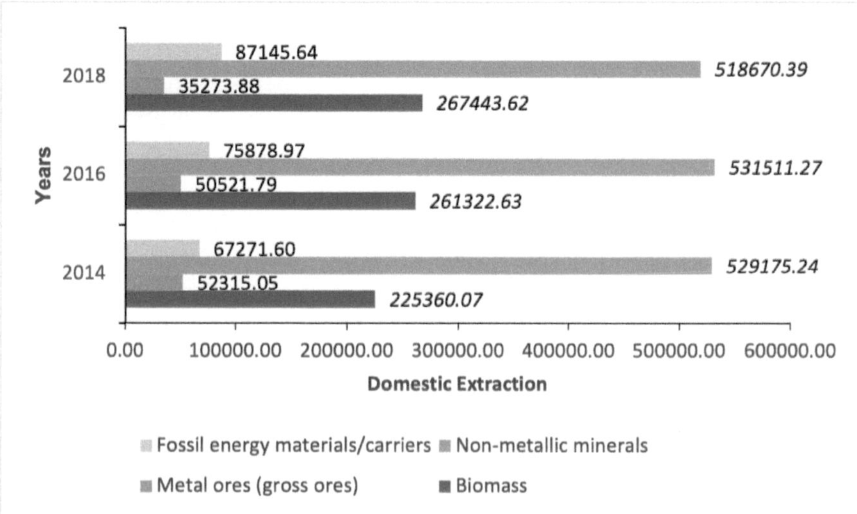

Figure 11.1. Domestic extraction by resource category (thousand tones).
Source: Data extracted from Eurostat (Eurostat, 2021b).

DMC represents the total amount of materials used within the economy and it is equal to domestic extraction plus physical imports minus physical exports. Figure 11.2 shows the DMC by resource category and indicates that while total DMC has increased by 10 percent in 2016, it has decreased by 3 percent in 2018 where much of the decrease is due to a

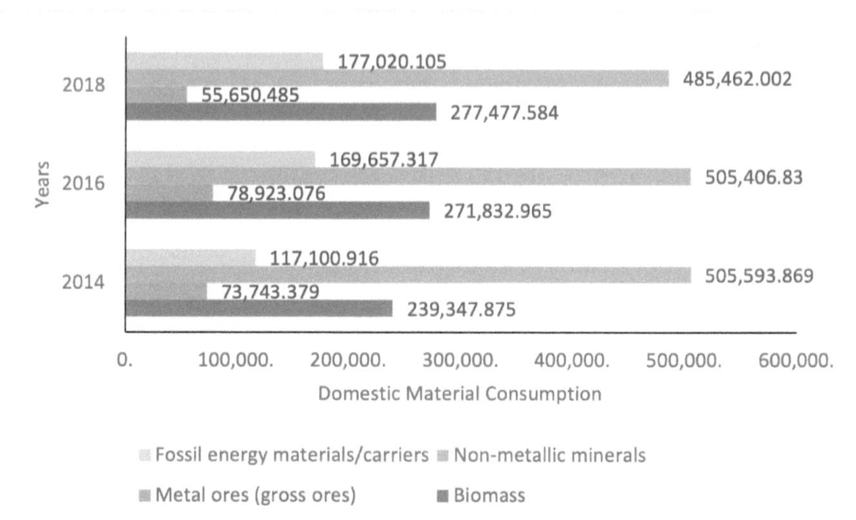

Figure 11.2. Domestic material consumption by resource category (thousand tones).
Source: Data extracted from Eurostat (Eurostat, 2021b).

29 percent decrease in metal ores consumption in 2018. Biomass consumption has been rising in Turkey by 14 percent and 2 percent and fossil energy materials/carriers is also increasing by 45 percent and 4 percent in 2016 and 2018, respectively. Non-metallic minerals consumption has been steady over the years.

Furthermore, comparing DE and DMC for each main material groups indicates that Turkish economy is dependent on foreign resources and is the net importer of all material groups except for the non-metallic minerals where it maintains its net exporter status.

Table 11.1 illustrates the four main material groups as a share of total material. Examining the data by material groups shows that non-metallic minerals as a percentage of total domestic materials extracted and consumed is higher relative to the other material groups for both DE and DMC, representing almost 50 percent of all materials. This higher rate of extraction and consumption of non-metallic materials which includes mostly construction materials such as sand, stone, granite, limestone, marble, and so on, is not surprising given the ever-expanding construction sector in Turkey. About 25 percent of DE and DMC consist of Biomass that represent crops, wood, timber, fish, and animals. Metal ores constitute the smallest group among main categories.

As the goal of CE is to keep the value of materials in circulation as long as possible, prevention of waste generation is vital in order to reach this goal. Analysis of the treatment of waste, by waste category [ENV_WASTRT] data set that represents the output side of the economy, that the amount of waste treated in Turkey has increased from 79,268 to 134,277 thousand tons, representing a 69 percent increase between 2014 and 2016 (Figure 11.3). However, data also reveals that almost 90 percent of waste treated in Turkey is sent to landfill causing high methane

Table 11.1. Shares of Materials as Percent of Total Materials

	2014		*2016*		*2018*	
Materials	*DE*	*DMC*	*DE*	*DMC*	*DE*	*DMC*
Biomass as percent of total materials	25.8%	25.7%	28.4%	26.6%	29.4%	28.1%
Metal ores (gross ores) as percent of total materials	6.0%	7.9%	5.5%	7.7%	3.9%	5.6%
Non-metallic minerals as percent of total materials	60.5%	54.2%	57.8%	49.4%	57.1%	49.1%
Fossil energy materials/ carriers as percent of total materials	7.7%	12.6%	8.3%	16.6%	9.6%	17.9%

Source: Data extracted from Eurostat (Eurostat, 2021b).

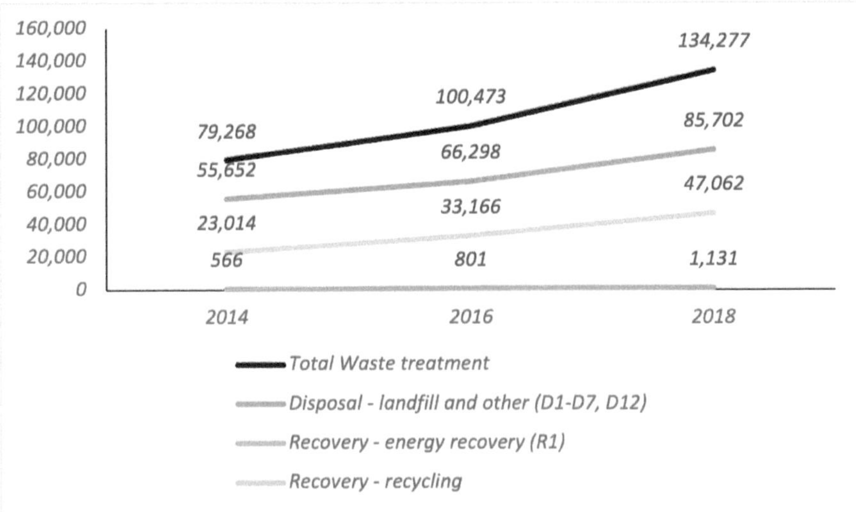

Figure 11.3. Waste Treatment Operations by Category (Thousand Tones).
Source: Data extracted from Eurostat (Eurostat, 2021c).

emissions in the country. The percentage of waste treated as incineration for energy recovery is less than 1 percent and rest has been recycled, as shown in Figure 11.3.

It is also relevant to analyze the recycling rate of municipal waste since it shows the percentage of waste generated by households is used as a resource in the CE. Figure 11.4 shows the recycling rate (R) in proportion

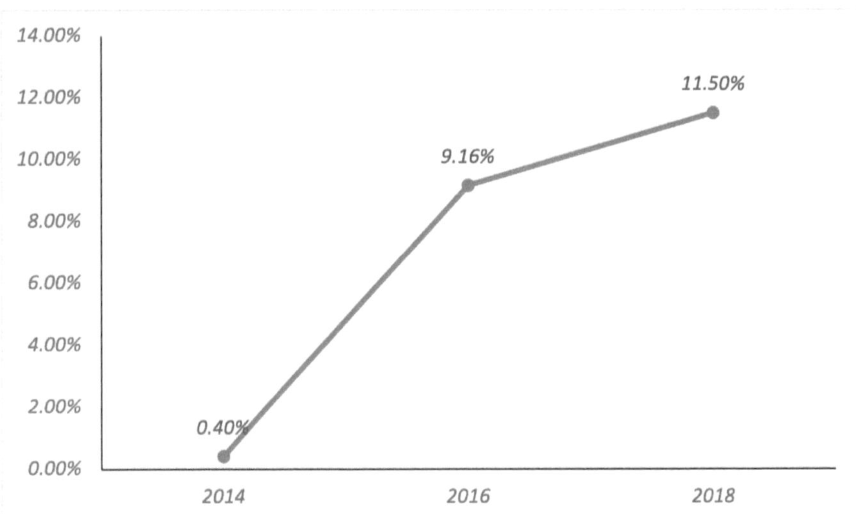

Figure 11.4. Recycling rate of municipal waste
Source: Eurostat, 2021e; Eurostat, 2021c.

of municipal waste generated that is recycled in Turkey has been increasing from 0.40 percent to 11.50 percent in 2018. However, compared to the target set by European commission at 65 percent by 2030 (European Commission, 2018), Turkey's recycling rates are quite low.

The CMU, the focus of this study, measures the ratio of recycled, secondary materials to the overall processed materials. Higher CMU represents that more recycled materials were able to replace the virgin raw materials, decreasing the burden on the environment. Calculations of recycled materials *(U)*, which is defined in equation 4, reveals 23,199 thousand tons, 33,358 thousand tons, and 47,177 thousand tons of secondary materials have entered back into the economy as recycled or downcycled in 2014, 2016, and 2018, respectively. Figure 11.5 indicates that the degree of circularity of Turkish economy measured as the share of recycled materials in total processed materials appears to be at 2.43 percent, 3.16 percent, and 4.56 percent in 2014, in 2016, and in 2018, respectively. The CMU rate for the Turkey has a steady upward trend from 2014 to 2018.

Although calculations of CMU present the increase in contributions of secondary materials into total processed materials by 30 percent in 2016, and 44 percent in 2018 relative to previous time periods; the improvement of CMU is partly due to decrease in Processed Materials (PM) (Figure 11.6), and partly due to improvement in the amount of waste recycled (Figure 11.4).

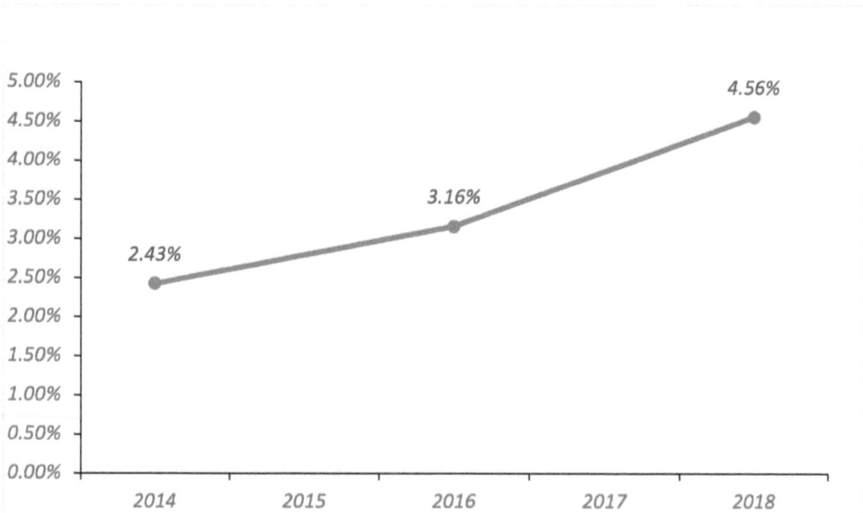

Figure 11.5. CMU in Turkey.
Source: Data extracted from Eurostat, 2021b; Eurostat, 2021c; Eurostat, 2021f.

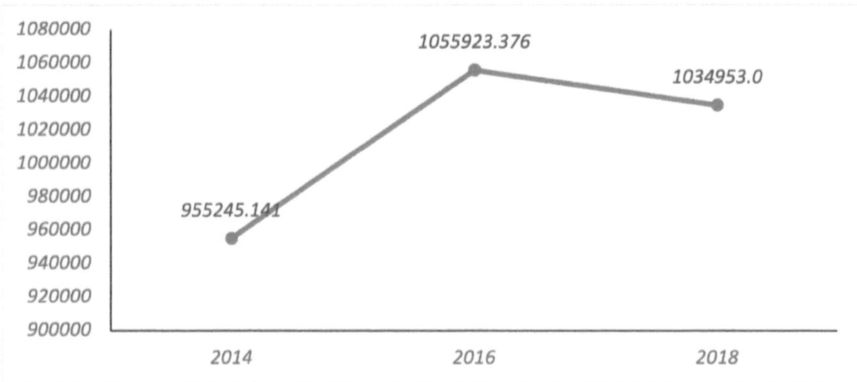

Figure 11.6. Processed materials in Turkey (Tones).
Source: Data extracted from Eurostat (Eurostat, 2021b).

Furthermore, comparing Turkey's CMU indicator with those European countries reveals that Turkey is lagging beyond in turning its economy form linear to CE. Figure 11.7 shows that in many major European countries, CMU rates are between 20 percent and 30 percent while Turkey's CMU rate is between 2.4 percent and 4.5 percent. This implies that Turkey would have to take serious steps to increase its circularity in order to catch up with other European countries as well as to decrease its dependence on primary materials and reduce the burden on environment.

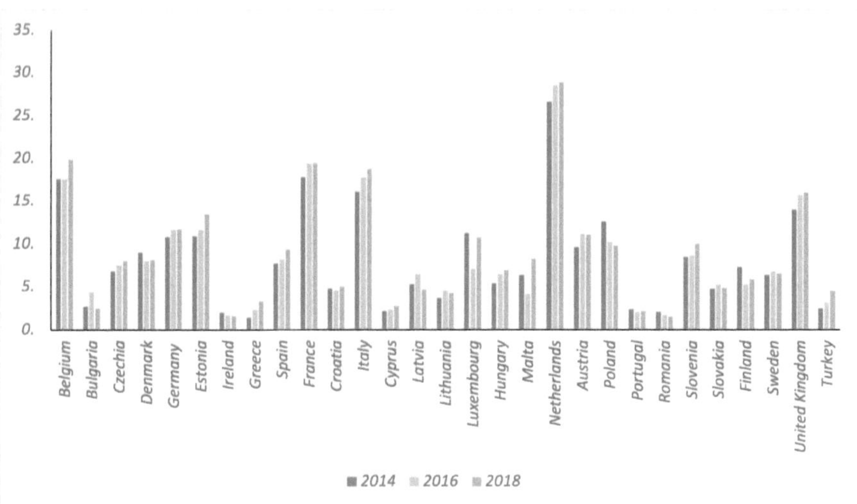

Figure 11.7. Comparison of CMU in Turkey with EU Countries.
Source: Data extracted from Eurostat 2021d.

It should be noted that conducting detailed Material Flow analysis and calculations of CMU for different material groups is necessary to be able to develop targeted policies to be able to increase circularity and decrease the use of primary resources in Turkey. For example, data on domestic material consumption reveals that about 50 percent of all materials consumed consists of non-metallic minerals that are mostly used in construction industry. Thus, it would be vital to design targeted policies to improve circularity for this group of materials. These include, but are not limited to, designing policies and programs to improve the existence of a market for recycled materials, to increase regional recycling capacities, and to educate labor and designer for using recycled materials in not only in construction projects. Different targeted policies for all sectors are essential in successfully transitioning into the CE.

Conclusion

The assessment presented in this article was an attempt to monitor the level of circularity of Turkey and track how it has evolved since 2014. The CMU rates showed that only 4.5 percent of materials processed in Turkey were recycled and contributed to closing the circularity loop in 2018. The results indicate that CMU of Turkey has an increasing trend between the years 2014 and 2018. Besides, CMU rate is growing at a high rate over the years, which is promising. However, the CMU rate is much lower than CMU rate of EU (twenty-eight) indicating that Turkey has much further to go if it is to keep up with EU member states. Currently, the results show CE is far beyond the sight for Turkey.

The analysis conducted in this study is not sufficient to assess the overall CE in Turkey. For a complete assessment of CE, input and output side circularity indicators should be analyzed in a holistic approach. CMU provides an information to have an insight of socio-economic loop closing (recycling waste as secondary material input) while in the scope of CE ecological loop closing (using renewable biomass) should also be evaluated. The slight increase in share of biomass in DMC indicates a possibility of growing ecological cycling potential for Turkey. However, it is not possible to make a proper assessment of ecological loop closing of Turkey due to the difficulty to reach confidential data and lack of data.

The implementation of policies enabling the transition toward CE is inevitable for Turkey due to the global tendency toward CE. Monitoring the success of the policies is dependent on the data availability of Turkey both on output and input side. Further research should be conducted to provide a more detailed dataset upon the reach of confidential data.

References

Allwood, J. M., Cullen, J. M., and Milford, R. L. (2010). Options for achieving a 50% cut in industrial carbon emissions by 2050. *Environmental Science & Technology*, *44*(6), 1888–94. https://doi.org/10.1021/es902909k.

Andersen, M. S. (2006). An introductory note on the environmental economics of the circular economy. *Sustainability Science*, *2*(1), 133–40. https://doi.org/10.1007/s11625-006-0013-6.

Arruda, E. H., Melatto, R. A. P. B., Levy, W., and Conti, D. De M. (2021). Circular economy: A brief literature review (2015–2020). *Sustainable Operations and Computers*, *2*, 79–86. https://doi.org/10.1016/j.susoc.2021.05.001.

Bennett, J. W. (1991). Pearce, D. W., and R. K. Turner. economics of natural resources and the environment. Baltimore md: Johns Hopkins university press, 1990, 378. *American Journal of Agricultural Economics*, *73*(1), 227–28. https://doi.org/10.2307/1242904.

Bocken, N. M. P., De Pauw, I., Bakker, C., and Van der Grinten, B. (2016). Product design and business model strategies for a circular economy. *Journal of Industrial and Production Engineering*, *33*(5), 308–20. https://doi.org/10.1080/21681015.2016.1172124.

Boulding, K. E. (1966). The economics of the coming spaceship Earth. In H. Jarrett (Ed.), *Environmental quality issues in a growing economy* (pp. 3–14). Baltimore, MD: Johns Hopkins University Press.

Bringezu, S. (2015). Possible target corridor for sustainable use of global material resources. *Resources*, *4*(1), 25–54. https://doi.org/10.3390/resources4010025.

Cullen, J. M. (2017). Circular economy: Theoretical benchmark or perpetual motion machine? *Journal of Industrial Ecology*, *21*(3), 483–86. https://doi.org/10.1111/jiec.12599.

EASAC—the European Academies' Science Advisory Council, 2016. *Indicators for a circular economy*. EASAC policy report 30. Halle (Saale), Germany: EASAC Secretariat, Deutsche Akademie der Naturforscher Leopoldina. Retrieved from: https://easac.eu/fileadmin/PDF_s/reports_statements/Circular_Economy/EASAC_Indicators_web_complete.pdf.

EMF, 2015. Towards the circular economy: Economic and business rationale for an accelerated transition. Cowes, UK: Ellen MacArthur Foundation. Retrieved from: https://www.werktrends.nl/app/uploads/2015/06/Rapport_McKinsey-Towards_A_Circular_Economy.pdf. Accessed on 27.09.2021.

European Commission, 2015. Communication from the Commission to the European Parliament, the Council, the European Economic and Social Committee and the Committee of the Regions—Closing the loop—An EU action plan for the Circular Economy COM (2015) 614/2 of 2 December 2015). Retrieved from: https://eur-lex.europa.eu/legal-content/EN/TXT/?uri=CELEX%3A52015DC0614.

European Commission, 2018. Commission staff working document—Measuring progress towards circular economy in the European Union—Key indicators for a monitoring framework. Accompanying the document: communication from the commission to the European parliament, the council, the European economic and social committee and the committee of the regions on a monitoring framework for the circular economy (COM (2018) 29 final, 19.1.2018). https://eur-lex.europa.eu/legal-content/EN/TXT/PDF/?uri=CELEX:52018SC0017&from=EN.

European Commission, 2019. Report from the Commission to the European Parliament, the Council, the European Economic and Social Committee and the Committee of the Regions on the implementation of the Circular Economy Action Plan, (COM (2019) 190 final,4.3.2019). https://eur-lex.europa.eu/legal-content/EN/TXT/?uri=CELEX%3A52019DC0190.

European Commission, 2020. Communication from the Commission to the European Parliament, the Council, the European Economic and Social Committee and the Committee of the Regions—A new Circular Economy Action Plan—For a cleaner and more competitive Europe, (COM (2020) 98 final,11.3.2020). https://eur-lex.europa.eu/legal-content/EN/TXT/HTML/?uri=CELEX:52020DC0098&from=EN.

European Commission. Statistical Office of the European Union. (2018a). *Circular material use rate: Calculation method: 2018 edition*. Publications Office. http://doi.org/10.2785/132630.

European Commission. Statistical Office of the European Union. (2018b). *Economy-wide material flow accounts handbook: 2018 edition*. Publications Office. http://doi.org/10.2785/158567.

European Environment Agency. (2016). *Circular economy in Europe: Developing the knowledge base*. Publications Office. https://doi.org/10.2800/51444.

Eurostat, 2021a. Statistics explained, Material flows in the circular economy. Retrieved from: https://ec.europa.eu/eurostat/statistics-explained/index.php?title=Material_flows_in_the_circular_economy. Accessed on 30/06/2021.

Eurostat, 2021b. Material flow accounts [ENV_AC_MFA]. Retrieved from: https://ec.europa.eu/eurostat/databrowser/view/env_ac_mfa/default/table?lang=en Accessed on 25/08/2021.

Eurostat, 2021c. *Treatment of waste-by-waste category, hazardousness and waste management operations* [ENV_WASTRT]. Retrieved from: https://ec.europa.eu/eurostat/databrowser/view/env_wastrt/default/table?lang=en. Accessed on 25/08/2021.

Eurostat, 2021d. Circular material use rate [ENV_AC_CUR]. Retrieved from: https://ec.europa.eu/eurostat/databrowser/view/sdg_12_41/default/table?lang=en. Accessed on 25/08/2021.

Eurostat, 2021e. Municipal waste by waste management operations [ENV_WASMUN]. Retrieved from: https://ec.europa.eu/eurostat/databrowser/view/env_wasmun/default/table?lang=en. Accessed on 25/08/2021.

Eurostat, 2021f. Transboundary shipments of notified waste by partner, hazardousness and waste management operations. Retrieved from: https://ec.europa.eu/eurostat/databrowser/view/env_wasship/default/table?lang=en. Accessed on 25/08/2021.

Fischer-kowalski, M., Krausmann, F., Giljum, S., Lutter, S., Mayer, A., Bringezu, S., Moriguchi, Y., Schütz, H., Schandl, H., and Weisz, H. (2011). Methodology and indicators of economy-wide material flow accounting. *Journal of Industrial Ecology*, *15*(6), 855–76. https://doi.org/10.1111/j.1530-9290.2011.00366.x.

Geisendorf, S., and Pietrulla, F. (2017). The circular economy and circular economic concepts-a literature analysis and redefinition. *Thunderbird International Business Review*, *60*(5), 771–82. https://doi.org/10.1002/tie.21924.

Geissdoerfer, M., Savaget, P., Bocken, N. M.p., and Hultink, E. J. (2017). The circular economy—A new sustainability paradigm? *Journal of Cleaner Production*, *143*, 757–68. https://doi.org/10.1016/j.jclepro.2016.12.048.

GEO5 (Global Environmental Outlook 5). 2012. Global Environmental Outlook 5—Environment for the future we want. ISBN: 978-92-807-3177-4. Nairobi: *United Nations Environment Program*. https://www.unep.org/resources/global-environment-outlook-5.

Ghisellini, P., Cialani, C., and Ulgiati, S. (2016). A review on circular economy: The expected transition to a balanced interplay of environmental and economic systems. *Journal of Cleaner Production*, *114*, 11–32. https://doi.org/10.1016/j.jclepro.2015.09.007.

Haas, W., Krausmann, F., Wiedenhofer, D., and Heinz, M. (2015). How circular is the global economy? An assessment of material flows, waste production, and recycling in the European union and the world in 2005. *Journal of Industrial Ecology*, *19*(5), 765–77. https://doi.org/10.1111/jiec.12244.

Jurgilevich, A., Birge, T., Kentala-lehtonen, J., Korhonen-kurki, K., Pietikäinen, J., Saikku, L., and Schösler, H. (2016). Transition towards circular economy in the food system. *Sustainability*, *8*(1), 69. https://doi.org/10.3390/su8010069.

Krausmann, F., Schandl, H., Eisenmenger, N., Giljum, S., and Jackson, T. (2017). Material flow accounting: Measuring global material use for sustainable development. *Annual Review of Environment and Resources*, *42*(1), 647–75. https://doi.org/10.1146/annurev-environ-102016-060726.

Lieder, M., and Rashid, A. (2016). Towards circular economy implementation: A comprehensive review in context of manufacturing industry. *Journal of Cleaner Production*, *115*, 36–51. https://doi.org/10.1016/j.jclepro.2015.12.042.

Mayer, A., Haas, W., Wiedenhofer, D., Krausmann, F., Nuss, P., and Blengini, G. A. (2018). Measuring progress towards a circular economy: A monitoring framework for economy-wide material loop closing in the eu28. *Journal of Industrial Ecology*, *23*(1), 62–76. https://doi.org/10.1111/jiec.12809.

Sapmaz Veral, E. (2019). Atık yönetiminde döngüsel ekonomi modeli ve döngülerin tasarımına İlişkin bir değerlendirme. *European Journal of Science and Technology*, 18–27. https://doi.org/10.31590/ejosat.479333.

Sapmaz Veral, E., and Hakan, H. (2018). Avrupa birliği atık politikasında atık yönetiminden kaynak yönetimi yaklaşımına geçiş yönelimleri ve döngüsel ekonomi modeli. *Ankara Üniversitesi Çevrebilimleri Dergisi*, *6*(1), 1–19. https://doi.org/10.1501/Csaum_0000000082.

Su, B., Heshmati, A., Geng, Y., and Yu, X. (2013). A review of the circular economy in China: Moving from rhetoric to implementation. *Journal of Cleaner Production*, *42*, 215–27. https://doi.org/10.1016/j.jclepro.2012.11.020.

T. C. Ticaret Bakanlığı, 2021. Yesil Mutabakat Eylem Planı. Retrieved from: https://ticaret.gov.tr/data/60f1200013b876eb28421b23/MUTABAKAT%20YE%C5%9E%C4%B0L.pdf Accessed on 27.09.2021.

Macro Indicators of Circularity in Turkish Economy

Opportunities and Challenges

12

AYŞE SEVENCAN AKCABELEN

Introduction

Environmental degradation and the "great acceleration" are among the world's main problems. The problem is more significant for developing countries with limited resources and growing populations; the need for circular business models is essential. Turkey is a net energy importer country. Among many other vital inputs required for production, the trade-off between increasing output for accelerating population and sustainable development creates an urgent need for a circular economy. Compared to industrialized countries, developing countries like Turkey have difficulties achieving Sustainable Development Goals (UNESCAP, 2021). Increasing population, predominantly urban areas without much improvements in urbanization in terms of establishing the required infrastructure for the high-density population, is leaving these countries at risk. Rapid economic growth in cities without the necessary infrastructure can significantly strain urban institutional resources and infrastructure in growing cities. In countries with lower levels of urbanization and higher growth, urban citizens are highly vulnerable—vulnerability refers to the lack of coping and urban adaptive capacities (Garschagen et al., 2016).

As mass manufacturing has developed over the last 150 years, mass production follows a linear system. The classical production cycle takes some materials, makes a product, and then discards it at the end of its use. This system of "take, make, waste" (the linear economy) has moved us into an ecological overshoot. Each year, the world population consumes 1.75 planets (Global Footprint Network, 2019). In the past fifty years, we have destroyed or degraded 60 percent of the earth's ecosystems—our life-support systems. To avoid the catastrophic impacts of climate change, global warming cannot exceed 1.5°C above pre-industrial temperatures (IPCC, 2018). That means we must halve greenhouse gas (GHG) emissions by 2030 and reduce them to net-zero by the middle of the century. The circular economy plays a significant role in reducing GHG emissions and supporting progress toward the United Nations Sustainable Development Goals (SDGs). The circulation of linear production chains into another process is closing the loop. As a result, materials that usually end up being landfilled are kept within the economy. In this way, the circular economy stimulates sustainable economic growth decoupled from extracting raw materials from nature. The Circular Economy Action Plan adopted in 2015 by the European Union already resulted in €147 billion in added value from circular processes such as recovering and recycling in 2016 (European Comission, 2019).

After covering the COVID-19 period, Turkey struggled with rising inflation and unemployment. In January 2022, the official consumer price inflation (CPI) of Turkey reached 48.7 percent (year-on-year), and domestic producer price inflation (PPI) hit 93.5 percent, the highest level in the last two decades. The sharp depreciation of the lira coupled with increased exchange rate pass-through, rising demand pressures, and deteriorating inflation expectations led to a jump in inflation. Food price inflation is the most significant contributor to headline inflation. In producer prices, energy prices, which are directly associated with oil prices, and even more with exchange rate movements, rose sharply in 2021 (World Bank, 2022a). Turkey's fragility in production costs calls for urgent policies on alternative energy sources and circular economy practices.

This chapter focuses on the main drivers of the circular economy in Turkey, namely, population, climate change, environmental degradation, and energy reliance. It lays out a picture of the country in terms of these drivers. Based on the current situation on drivers for a change for a more circular economy in Turkey, the chapter represents the possible future for the country. Part 2 of this chapter discusses the drivers for the circular economy in Turkey; part 3 outlines the possible accelerators and policies that may lead to a more circular economy, and the last part concludes.

Drivers for Circular Economy in Turkey

According to the OECD Survey on the Circular Economic in Cities and Regions, climate change, global agendas, and economic changes are significant drivers for surveyed cities and regions to transition to a circular economy (OECD Survey, 2020). Major drivers for transitioning to a circular economy are environmental (climate change, 73 percent), institutional (global agendas, 52 percent), and socio-economic (changing economic conditions, 51 percent). Additionally, the circular transition is driven by job creation (47 percent), private sector initiatives (46 percent), new business models (43 percent), technical developments (43 percent), and research and development (R&D) (41 percent). The keywords respondents most associates with the circular economy in cities and regions are "climate change," "zero waste," and "innovation" (OECD Survey, 2020).

Population

One of the main drivers of the Circular Economy in Turkey is the growing population. In 1960, total population was around twenty-seven million, and in the last sixty years, it has tripled to eighty-four million in 2020 (World Bank, 2022). If the country's population increases at the same rate, the population will be above 100 million by 2040. This puts further pressure on land use and requires the construction of new homes and infrastructure, including water and sewage pipelines, railways and roads, and the provision of energy and other services. The population density in Turkey was around thirty-seven people per square kilometer of land area, and by 2020, it would be almost 110 people per square kilometer of land area. (World Development Indicators, 2022). Again, with a similar increase in population by 2040, the same area density of people would be around 132. Istanbul Bursa region of Turkey was the sixteenth mega-city region globally, with over fourteen million residents and a total GDP of 626 billion US dollars (Florida, 2019).

Environmental Degradation

Turkey has been experiencing environmental pressures due to population growth, industrialization, and rapid urbanization. These pressures translate into various ecological challenges such as climate change, desertification, deforestation, water scarcity, natural degradation, and marine pollution. Turkey has adopted new legislation and institutional practices to comply with the EU environmental acquis to address these challenges. Enforcement of environmental law began in the early 1980s, since which time there has been increasing regulation in the ecological sphere. According to the last report of the European Commission on Turkey, the adaptation of EU regulations in Turkey has led to improvements in several environmental indicators. Some of these improvements are a 29 percent decrease in air pollution from PM_{10}[1] and a 43 percent decrease in air pollution from sulfur dioxide (SO2) between 2007 and 2010. However, in just two years (between 2008 and 2010), the municipal waste collection increased by 3.8 percent. Regarding waste sites in urban areas in Turkey, the number of landfills, which was fifteen by 2001, rose to thirty-eight in 2008, fifty-nine in 2011, and sixty-nine in 2012. Between 2009 and 2010, there was a 25 percent increase in the amount of hazardous waste treated (European Commission, 2016).

The ecological footprint measures how much area of biologically productive land and water an individual, population, or activity requires to produce all the resources it consumes and absorb the waste it generates, using prevailing technology and resource management practices (Global Footprint Network, 2022). In 2018, Turkey's per capita Ecological Footprint of consumption was 3.4 gha, higher than the global footprint (2.8 gha) and exceeding the global biocapacity per capita (1.3 gha)[2] by 260 percent. In other words, if every individual on earth would consume as much as an average Turkish citizen, humanity would need 2.6 planets. As Turkey consumes its natural resources faster than the speed they can renew themselves, since 1981, it has been running a biocapacity deficit (Global footprint network, 2022). The main reason Turkey transformed from a country with biocapacity reserve to a country with a biocapacity deficit is the acceleration of population.

According to the Turkish Statistical Institute database, the greenhouse gas inventory results revealed that overall greenhouse gas (GHG) emissions as CO_2 equivalent (eq.) for the year 2020 compared to the previous year increased by 3.1 percent to 523.9 million tons (Mt). Total GHG emissions per capita were calculated at four tons CO_2 eq. for 1990, 6.2 tons CO_2 eq. for 2019, and 6.3 tons CO_2 eq. for 2020 (Turkstat, 2022).

Energy

Turkey has a strong energy demand and depends on imported energy because of rapid economic and population growth in the last two decades (OECD, 2022). Although there is an increase in the installed capacity of renewable sources, the share of renewable energy in the total energy supply is meager. In twenty years, the percentage of renewable energy in the total energy supply has increased by only 3 percent (from 13.2 percent in 2000 to 16.2 percent in 2020) (IEA, 2022).

Energy is one of Turkey's leading contributors to air pollution; in 2020, the energy sector had the largest share of total GHG emissions, with 70.2 percent. The agriculture sector followed the energy sector with 14 percent, the industrial processes and product use sector with 12.7 percent, and the waste sector with 3.1 percent. Renewable energy use is more promising in electricity production, however. The share of renewables has risen from 25 percent to 42 percent between 2000 and 2020 (IEA,2022)

The energy sector emissions were calculated at 367.6 Mt CO_2 eq. in 2020, which increased by 163.3 percent compared to 1990 and by 0.6 percent compared to the previous year. Similarly, emissions from the industrial processes and product use sector were calculated at 66.8 Mt CO_2 eq. in 2020, which increased by 190.5 percent compared to 1990 and 14 percent compared to the previous year (Turkstat, 2022). 85.4 percent of total CO_2 emissions originated from the energy sector, including 31.6 percent of total CO_2 emissions from electricity and heat production, a sub-category of the energy sector. The remaining 14.2 percent of CO_2 emissions originated from the industrial processes and product use sector and 0.4 percent from the agriculture and waste sectors in 2020. Sixty-one percent of CH_4 emissions originated from agriculture, 22.1 percent from waste, 16.9 percent from energy, and 0.02 percent from industrial processes and product use. In comparison, 80.3 percent of N_2O emissions originated from the agriculture sector, 9.1 percent from energy, 5.6 percent from waste, and 5 percent from industrial processes and product use (Turkstat, 2022).

Turkey's economic progress is primarily reliant on fossil fuel-based solutions, and this reliance on fossil fuels casts doubt on the ecological sustainability of the present economic expansion trend (He et al., 2021). Turkey is experiencing difficulties reconfiguring its current policies to achieve the goal of climate action, which is covered under Sustainable Development goals (SDG) 13.

Climate Change

The UN Intergovernmental Panel on Climate Change (IPCC) report in 2021 summarized the threats to Turkey in three headlines: rising temperatures, dehydration, and rising sea levels (IPCC, 2021). By 2050, temperatures are predicted to increase by 2.5° Celsius in the east and central Turkey and by 1.5° Celsius on the coasts. Temperatures exceeding 40° Celsius are expected in the summer for extended periods. Annual precipitation is expected to gradually fall by around 10 percent by 2050, resulting in increased water scarcity and periods of drought, problems exacerbated by glacial retreat, and decreased snowfall in the mountains, from which half of the country's water is derived.

With the limited water supply by nature, Turkey faces problems regarding efficient management of water resources, just like energy. Population growth, urbanization, and expansion of irrigation areas generate higher water demand and increase pressures on water resources. Total water abstractions have increased since 2007, putting the country under medium water stress, which implies the management of both supply and demand, and conflicts among competing uses need to be resolved (OECD, 2022b). The abstractions are decreasing in countries like Germany, which is an example of the successful implementation of environmental policy. Germany reduced total water abstractions from 32.5 percent of the internal resources in 2001 to 20.9 percent in 2016. On the other hand, in Turkey, total water abstractions have increased from 19 percent to 26.9 percent between 2000 and 2016.

Agriculture is the sector that uses the most abstracted freshwater. Water scarcity will also lead to greater food insecurity. In 2018, OECD reported that the country's 74 percent of the total water consumption is agricultural water consumption (OECD, 2020). Inevitably, strict regulations will be enacted for the farming and animal husbandry sectors. The National Water Plan of Turkey's policy is five years between 2019 and 2023. It aims to ensure efficient, rational, and sustainable use of existing water resources before finding new water resources and offering them to the user (Gruère et al., 2020).

The Need for Sustainable Economic Growth

Using resources efficiently is the pillar of sustainable economic growth. Turkey needs sustainable economic development to meet the needs of its increasing population. As mentioned in this chapter, energy, and water are the two natural resources Turkey needs to use efficiently. Another resource is material consumption. Efficiency requires minimum use of resources for the maximum amount of production. Circular economy takes the efficiency requirement as a baseline of sustainable production.

There are two ways to measure material consumption in economies. The first one is the domestic material consumption (DMC), which is an indicator that measures the number of materials directly used in an economy. It is defined as the domestic extraction of materials plus all physical imports minus all physical exports (OECD, 2022c). Second, a complementary indicator is a material footprint, also known as raw material consumption (RMC), which represents the final domestic use of products in terms of raw material equivalents. It captures the amount of domestic and foreign extraction of materials needed to produce the final products used domestically and is defined as domestic extraction of materials, plus imports in raw material equivalents, minus exports in raw material equivalents (OECD, 2020).

In 1996, DMC was 10.3 tons per capita in Turkey, was the lowest in 2001 crisis at 8.2 techniques per capita, peaked at 13.3 tons per capita in 2011 and 10.6 tons per capita, compared to OECD average and industrialized EU such as Germany, the country has lower domestic material consumption per capita, for example, similar population Germany DMC was 14.2 tons per capita in 2019 (OECD, 2020).

Turkey is endowed with natural resources, such as antimony, coal, chromium, mercury, copper, borate, sulfur, and iron ore. Thanks to high economic growth, material productivity increased over the past decade but remained below the OECD average. Domestic material consumption (DMC) intensity per capita peaked in 2011 and decreased afterward, despite a rebound in 2017. Turkey's material footprint intensity grew faster than DMC intensity since 2006, reflecting the increase in imports of intermediate goods. Material productivity is expressed as the economic output generated (GDP) per unit of materials consumed (DMC). This indicator is measured in USD constant prices using the 2010 base year and Purchasing Power Parities (PPPs) (OECD, 2022a). One might be cautious about interpreting the productivity accounts of Turkey. Considering the high level of inflation and deterioration of the Turkish Lira since 2010, material productivity should be much lower in the 2021 exchange rate. The PPP has increased from 0.90 in 2010 to 2.6 in 2021. The yearly exchange rate has increased from 1.5 to 8.6 during the same period (OECD, 2022d).

Material productivity is increasing but not converging with developed nations; in other words, the rate of increase is not as much as in developed countries. In 2019, 2.6 USD per kilogram in Turkey, whereas in the UK, Germany, and Europe, material productivities are 5.4, 3.5, and 3.6, respectively.

As in many other OECD countries, in Turkey also, construction minerals constitute the bulk of the materials used. The share of fossil energy carriers, especially coal, lignite, oil, and natural gas, increased over the past decade (OECD, 2022c).

Policies Contributing to Circular Economy

Circular economy principles guarantee a sustainable world and better economic opportunities for countries. Reusing materials and closing the loops can decrease production costs and create new job opportunities (Murray et al., 2017). Policies should focus on creating awareness on the subject from an environmental perspective and cost efficiency and profit opportunities for contributing industries to circularity.

Turkey is already elevating climate and environmental policy, and linking this plan to the recovery program could accelerate progress. A strong focus on cross-sectoral policies that stimulate technological innovations and modernize the economy can help Turkey improve material efficiency, maintain competitiveness, create jobs, and support overall development priorities. The stimulus used in this way would not cost more—but instead be used more efficiently and for more sustainable outcomes. The path to greener technology adoption in many sectors is quite established, and steps could be taken to align with the evolving requirements of the EU and other key trade partners. Stimulus resources could incentivize resource efficiency, and Turkish position firms strategically if such instruments as the Carbon Border Adjustment Mechanism (CBAM) come to fruition. A wide range of tools can stimulate resource efficiency and improve environmental performance. These instruments can be categorized into three groups. Regulatory agencies that mandate specific behavior; market-based instruments that act as incentives for particular activities; and information-based instruments seek to change behavior through the provision of information (World Bank 2022).

In a worldwide study, Dellink (2020) explores how the widespread implementation of circular economy policies may impact trade patterns globally. This study analyses the consequences of a stylized package of circular economy policies on trade patterns to 2040. Results show that, compared to a scenario with no additional circular economy policies, the policy package leads to more substitution of primary with secondary raw materials, although the secondary sector remains small (Dellink, 2020). Implementing circular economy policies in countries like Turkey, would increase the comparative advantage in secondary materials.

Turkey has several opportunities to utilize its current stimulus package more targeted ways. Elements of Turkey's stimulus package are considered green, including introducing a "Green Tariff" for power generated from renewable sources, support for solar power, and extension of the Renewable Energy Support Scheme. Turkey is also committed to increasing solar energy production capacity by 1 GW, while the recently announced ERP includes commitments to support further green reforms (World Bank, 2022b).

Unfortunately, public awareness regarding climate change and environmental degradation is low in Turkey. The country suffers from high inflation and rising costs due to the deterioration of the Turkish Lira and increases in oil prices. Inflation along with unemployment has been the dual problem of the country recently. In the last Future of Business Survey in January 2022, only 24 percent of the small business in Turkey are concerned with the costs of implementing climate change policies and regulations in the coming fiscal year.

Waste Management

Circular economy requires resources efficiency, and materials are used efficiently during the production process and at all stages of their life cycle. The material lifecycle also covers the period after consumption, either recovered or disposed of. The OECD and Ellen Mac Arthur foundation define circular economy policies as a set of reduce reuse and recycling stages and waste prevention. In the heart of the circular economy, nothing should be considered as "waste," either reused, composted, remanufactured, or reimbursed (Ellen Mac Arthur Foundation, 2013; OECD, 2019).

The OECD database on Waste Management shows some critical insights for Turkey. Although Turkey's waste creation is lower than the OECD average, Turkey lags behind its industrialized counterparts in disposing of waste environmentally (OECD, 2022b). OECD takes municipal waste as an indicator of how waste, in general, is treated within the economy. As stated by OECD, although waste collected by or on behalf of municipalities is only a small share of total waste generated (about 10 percent), its management and treatment often represent more than one-third of public sector financial efforts to abate and control pollution (OECD, 2022b). The least environmental way of waste management is landfilling, a deposit of waste into or onto land, including specially engineered landfills and temporary storage of over one year on permanent sites. Other ways are recycling, composting, and incineration with or without energy.

The generation of municipal waste per capita in Turkey is below the OECD average. However, 87.7 percent of the total municipal waste was sent to landfills in 2019, the highest share in OECD. In a country where scarce water resources are mainly used for agriculture, unfortunately, only 0.04 percent of the municipal waste was composted in 2019, the lowest among the OECD members. For the same year, 29.2 percent of the municipal waste was composted in the Netherlands. The share of recycled municipal waste is again lower than most OECD countries; only 11.2 percent of the waste is recycled (OECD,2022b). Recycling has gained importance since 2017 with the "Sıfır Atık" (Zero Waste) Project (Zero Waste, 2022).

To examine the trend of sustainable economic growth, looking at the waste management pattern of a country over a time period is also helpful. Although the amount of total waste has been increasing in Turkey for the last two decades due to increased population and urbanization, unlike the industrialized members, composting and incineration practices are not increasing in Turkey. Actually, in a country that heavily depends on imported fuel energy, the share of incineration is zero (OECD, 2022b).

Energy Policy

A circular economy requires any resource used in the production process to be in one way or another to be used in the process of production. In that sense, renewable energy use is vital for a circular economy. Between 2000 and 2020, Turkey has increased the diversity of its energy sources by increasing renewable energy use, especially in electricity consumption. The commissioning of Turkey's first nuclear power facility in 2023 will further diversify the country's energy mix (IEA, 2022). However, according to OECD (2022b), the government's tax exemptions implicitly support fossil fuel consumption in industry sectors. Further, the absence of a specific carbon tax and reliance on energy taxes do not price the number of carbon emissions accordingly. Less than 40 percent of CO_2 emissions are priced, and only 2 percent are priced above 120 EUR/ton of CO_2(IEA,2022).

In their last report, IEA (2022) recommends that Turkey create infrastructure and incentives to finance energy efficiency projects, define mid-term and long-term policies for developing renewable energy, and plan for emission reduction and air pollution targets to help sustainable energy policymaking (IEA, 2022).

The increasing demand for energy in the last sixty years is one of the Turkish economy's significant challenges. Aside from the environmental damage, increasing demand for imported energy leaves the country at risk in terms of sustainable development. Turkey's rapid economic and population growth has strongly driven up energy demand and increased import dependency in the past two decades. As a result, Turkey has pursued a restructuring of its energy system to rationalize energy demand growth, lower energy prices for consumers, and slow the pace of import growth (World Bank 2022b).

Due to its high dependence on oil and gas imports, Turkey has prioritized the security of its energy supply as one of the pillars of its energy strategy. The policy includes boosting domestic oil and gas exploration and production, diversifying oil and gas supply sources and associated infrastructure, increasing renewable energy production, and improving energy efficiency. Turkey has made particular progress in expanding its gas supply options through discoveries, pipelines, LNG terminals, and increased storage (IEA, 2022).

Environmental Label Project of Turkish Ministry of Environment and Urbanization

Turkey is taking steps to establish sustainable supply chains and increase its competitiveness in circular trade. While the regulatory restrictions on non-green imports for the European Union, Turkey needs to take the necessary steps as a committed trade partner. The environmental label project of the Turkish Ministry of Environment and Urbanization started in 2013 to develop a basis for a voluntary environmental label system in Tırkey in line with the EU Ecolabel Directive. The system set-up and legislation, logo design, regulatory impact assessment, training activities, software needs assessment, publicity of environmental label, and organization of seminars were performed in the project's scope. The draft regulation, national environmental label logo, regulatory impact assessment, software feasibility report, and publicity action plan were prepared due to the project. The primary purpose of the Environmental Labelling scheme is to support sustainable production and consumption patterns by encouraging organizations that have been carrying out their operations in a sensitive manner to the environment. The Eco-Label considers areas such as low carbon emission, waste prevention, energy efficiency, water-saving, and the use of non-hazardous chemicals (Turkish Ministry of Environment and Urbanization, 2022).

Conclusion

Turkey is a developing country in the southeastern belt of the Mediterranean, with a limited water supply, and depends heavily on imported energy. High population growth puts pressure on the country's limited resources and requires even higher economic growth. These characteristics of Turkey are similar to any developing country. Sustainable economic development is crucial for developing countries, and yet they lag their developed counterparts in terms of acting on switching to more ecologically and economically sustainable solutions.

This chapter summarized Turkey's main drivers for switching to a more circular economy: population growth, environmental deterioration, climate change, water scarcity, energy dependence, and material use. Switching to renewable energy sources is crucial for lowering costs and sustainable economic growth. The population is not distributed evenly within the country; accelerating urbanization is alarming. Waste management is far behind the industrialized countries in Turkey. Although much of the scarce water is used by the agricultural sector within the country, composting is not used as a method for waste management. For developing countries like Turkey, a circular economy is more essential than in industrialized countries, and policies should focus on creating opportunities for more circular applications.

Notes

1. PM_{10} describes inhalable particles, with diameters that are generally 10 micrometers and smaller.

2. Biocapacity: There were ~12 billion hectares of biologically productive land and water areas on Earth in 2017. Dividing by the number of people alive in that year (7.5 billion) gives 1.6 global hectares per person. This area also needs to accommodate the wild species that compete for the same biological material and spaces as humans (Global Footprint Network, 2022).

References

Dellink, R. (2020), "The consequences of a more resource-efficient and circular economy for international trade patterns: A modeling assessment," OECD Environment Working Papers, No. 165, OECD Publishing, Paris, https://doi.org/10.1787/fa01b672-en.

Ellen MacArthur Foundation (2013), Towards a Circular Economy—Economic and Business Rationale for an Accelerated Transition. Cowes, Isle of Wight.

Ellen MacArthur Foundation and the McKinsey Center for Business and Environment (2015), Growth within A circular economy vision for a competitive Europe. Dublin, Ireland: Foundation for Environmental Economics and Sustainability.

European Commission (2019), "Closing the loop: Commission delivers on Circular Economy Action Plan," https://ec.europa.eu/commission/presscorner/detail/en/IP_19_1480.

Florida, R. (2019) "The real powerhouses that drive the world's economy," CityLab, 28 February, https://www.citylab.com/life/2019/02/global-megaregions-economicpowerhouse-megalopolis/583729/, last accessed March 8, 2022.

Garschagen, M., Hagenlocher, M., Comes, M., Dubbert, M., Sabelfeld, R., Lee, Y. J., Grunewald, L., Lanzendörfer, M., Mucke, P., Neuschäfer, O., Pott, S., Post, J., Schramm, S., Schumann-Bölsche, D., Vandemeulebroecke, B., Welle, T., Birkmann, J., (2016). World Risk Report 2016. Bündnis Entwicklung Hilft and UNU-EHS.

Gruère, G., M. Shigemitsu and S. Crawford (2020), "Agriculture and water policy changes: Stocktaking and alignment with OECD and G20 recommendations," *OECD Food, Agriculture and Fisheries Papers*, No. 144, OECD Publishing, Paris. http://dx.doi.org/10.1787/f35e64af-en.

He, K., Ramzan, M., Awosusi, A. A., Ahmed, Z., Ahmad, M., and Altuntaş, M. (2021), "Does globalization moderate the effect of economic complexity on CO2 emissions? Evidence from the top 10 energy transition economies.," *Frontiers in Environmental Science*, 555.

IEA (2022), "World energy statistics," *IEA World Energy Statistics and Balances* (database), https://doi.org/10.1787/data-00510-en (Accessed on March 24, 2022).

IPCC (2018), "Summary for Policymakers.," in Global Warming of 1.5°C. An IPCC Special Report on the impacts of global warming of 1.5°C above pre-industrial levels and related global greenhouse gas emission pathways, in the context of strengthening the global response to the threat of climate change, sustainable development, and efforts to eradicate poverty [Masson-Delmotte, V., P. Zhai, H.-O. Pörtner, D. Roberts, J. Skea, P. R. Shukla, A. Pirani, W. Moufouma-Okia, C. Péan, R. Pidcock, S. Connors, J. B. R. Matthews, Y. Chen, X. Zhou, M. I. Gomis, E. Lonnoy, T. Maycock, M. Tignor, and T. Waterfield (eds.)]. World Meteorological Organization, Geneva, Switzerland, 32 pp.

Murray, A., Skene, K., and Haynes, K., (2017) "The Circular Economy: An Interdisciplinary Exploration of the Concept and Application in a Global Context.," *Journal of Business Ethics*, *140*(3). pp. 369–80. ISSN 0167-4544.

OECD (2019), Global Material Resources Outlook to 2060: Economic Drivers and Environmental Consequences, OECD Publishing, Paris, https://doi.org/10.1787/9789264307452-en.

OECD (2020), Agricultural Policy Monitoring and Evaluation 2020, OECD Publishing, Paris, https://doi.org/10.1787/928181a8-en.

OECD (2022a), "Turkey," in *Environment at a Glance Indicators*, OECD Publishing, Paris, https://doi.org/10.1787/98471d84-en.

OECD (2022b), "Water: Freshwater abstractions," OECD Environment Statistics (database), https://doi.org/10.1787/data-00603-en.

OECD (2022c), Material consumption (indicator). doi: 10.1787/84971620-en (Accessed on 10 March 2022).

OECD (2022d), Exchange rates (indicator). doi: 10.1787/037ed317-en (Accessed on 24 March 2022).

World Bank (2022a), "Turkey Economic Monitor February 2022: Sailing Against the Tide." Turkey Economic Monitor; Washington, DC: World Bank. © World Bank.https://openknowledge.worldbank.org/handle/10986/37035 License: CC BY 3.0 IGO.

World Bank (2022b), Unlocking Green Finance in Turkey, Washington, DC: World Bank, 2022.

World Development Indicators (2021), World Bank Agriculture, forestry, and fishing, value added (percent of GDP) (2021) Washington, DC: The World Bank.

World Economic Forum. (2014, January). Towards the circular economy: Accelerating the scale-up across global supply chains. Retrieved August 2, 2015, from: http://www3.weforum.org/docs/WEF_ENV_TowardsCircularEconomy_Report_2014.pdf.

Turkish Statistical Institute (2022), https://data.tuik.gov.tr/Bulten/Index?p=Greenhouse-Gas-Emissions-Statistics-1990-2020-45862 (Accessed on March 24, 2022).

Turkish Ministry of Environment, Urbanization and Climate Change (2022), Environmental Label Project, https://cevreetiketi.csb.gov.tr/en/establishment-the-basis-of-environmental-label-project-i-100390, (Accessed on March 14, 2022).

York University Ecological Footprint Initiative and Global Footprint Network. National Footprint and Biocapacity Accounts, 2022 edition. Retrieved [March 20, 2022] from: https://data.footprintnetwork.org.

Zero Waste (2022), http://zerowaste.gov.tr/en/zero-waste/what-is-zero-waste, (Accessed on March 14, 2022).

An Evaluation of Circular Economy Post COVID-19

The Case of Turkey

13

DUAA BARAKAT AND ALPER ALTINANAHTAR

Introduction

Considering the current world, where rising industrialization and resource use are threatening the ecosystem in which we live, efficient resource use, pollution prevention, and reduction at the source are critical. Governments, businesses, and communities are seeking for methods to use the circular economy to decrease waste and the harmful consequences on society. Every year, more than 100 billion tons of resources are transferred to the economy, yet more than 60 percent of them are wasted, resulting in a rise in greenhouse gas emissions, one of the causes of the climate catastrophe (KPMG, 2022).

Turkey's economy has been significantly harmed by the COVID-19 pandemic, and its institutions lack a clear circularity understanding. A national circular economy plan has not yet been developed in Turkey; nevertheless, the "Green Deal Action Plan" was launched in July 2021 by the Ministry of Industry and Trade. The ministry has studied the adoption of circularity principles in Turkey (KPMG, 2022). In the first half of 2020, the COVID-19 pandemic halted progress toward the Sustainable Development Goals in Turkey and many other nations throughout the world, and it is expected to have an impact on the 2030 Agenda (UN Turkey, 2020). The crisis has had particularly apparent implications for populations and reinforced the relevance of the idea of "leaving no-one behind" (UN Sustainable Development Group, 2022). A COVID-19 recovery approach shows how governments might help pave the way for such a future of investment possibilities across important sectors as healthy environment, logistics, plastic packaging, recycling, and renewable energy. The circular economy (CE) is one of these approaches. The CE concept's major goal is to take advantage of material flow through recycling, strike a balance between economic growth and development, green environment, and resource consumption. The concept of CE is now more broadly accepted, and organizations around the world, such as the European Commission and the Ellen MacArthur Foundation, are supporting specific components of it, such as material design and flow evaluation. The resource products-pollution mode is one of the conceptual frameworks proposed by Pearce and Turner for the CE idea. Reducing, reusing, and recycling are the three principles of the CE idea (Andersen, 2006).

The "Zero Waste Project" aims to maximize resource efficiency while also preventing or reducing trash and garbage formation by collecting and reusing the smallest amount of waste at the source (UN Turkey, 2020). Industrial symbiosis, renewable energy use, digital design in production for raw material efficiency, product designs to extend product life, and energy efficiency by method engineering in supply chains are the circular economy implementation Systems. This chapter develops a framework to identify the implementation of circular economy models (reuse, recycling, sharing) by Turkish corporations post-COVID-19 pandemic. One of Turkey's post-pandemic goals is to break out the recession by encouraging the efficient use of resources. Increased revenue from new circular activities, along with lower production costs achieved by making items and materials more functional and readily disassembled and reused, has the potential to boost GDP and thus economic growth. Businesses that adopt the circular economy model can save input costs and create new profitable opportunities. Profit potential in this circularity may arise from entering new markets, reducing costs through waste and energy reductions, and ensuring supply continuity. On one hand, the prevention of the exposure to hazardous chemicals is a direct effect of the circular economy, whereas the indirect effect is a change in air quality due to reduced emissions from production and consumption. On the other hand, manufacturing shortages, disrupted transportation systems, international trade dynamics, and price shocks are all threats to linear supply-chain systems. These flaws were recognized during the COVID-19 pandemic, revealing that while single-use disposables seem harmless in normal circumstances, relying on them in extreme cases like the COVID-19 pandemic, raises the chance of catastrophic failure (CIF, 2021).

Furthermore, price and quantity controls, sales tariffs and unlicensed electricity generation permits, emission taxes, development of smart technologies, R&D support, and financial and economic aids for renewable energy are just a few of the policy instruments available to Turkish

Government to reduce fossil fuel consumption and encourage the use of renewable energy sources. Investments in renewable energy resources will increase as a result of these policy measures, while the usage of fossil fuel energy sources will decline. As a result, pollution will be reduced, and a sustainable ecology will emerge. According to Kilickaplan and others (2017), electricity produced with photovoltaic technology, PV, will be the greatest contribution to meeting Turkey's rising energy demand, accounting for 43.2 percent of total demand by 2050. There are few studies on the relevance of solar energy for Turkey's circular economy. The impact of the EU Circular Economy Action Plan on Turkey's energy policy and renewable energy investments was discussed. As a result, the Sunrise Initiative was chosen by the European Commission at the start of 2019 and is being financed by Horizon 2020 (Abbott, 2019). The Sunrise Initiative, titled "Solar Energy for a Circular Economy" is supported by Mersin University in Turkey with the concept of recycling and zero waste projects. Table 13.1 shows waste data from the Evaluable Waste Materials Industrialists' Association (Tüdam) (2021), which was used to compare waste import and export statistics from EU nations including Turkey in 2020.

The chapter aims to summarize the direct and indirect efforts by Turkey towards establishing circular economy components through reusing various medical devices, altering hospitals and companies' supply chains, recycling of chemicals, switching to clean energy, and reducing air pollution to enhance public health.

The chapter will continue as follows; there will be the literature review of circular economy and some examples of its worldwide applications. The third section reveals information about the organizations, in charge of applying circular economy concepts in Turkey. The forth section consists of Turkish government and associations' practices towards sustainability post COVID-19 pandemic. The last section will cover the conclusion and some advice for the government to enhance the sustainability in Turkey.

Literature Review

Erat and Telli (2020), using the Kao co-integration technique—the one-way causality interaction between hydroelectricity energy, renewable energy consumption, and nonrenewable energy consumption on economic growth in the E7 countries including Turkey—investigated and showed that these countries are the fastest-growing energy-dependent economies. Moreover, Barbier and others (2020) found that in the United States, the way toward circular economy is through fossil fuel subsidy swap to support clean energy projects and renewable energy diffusion. They also emphasize

Table 13.1. Comparison of EU Countries Waste Data (Tüdam, 2021)

Country	*Population (millions)*	*Packaging Waste (%)*	*Formed Packaging Waste (million ton)*	*Collected Packaging Waste (million ton)*	*Waste Imports (million €)*	*Waste Export (million €)*	*Foreign Trade Gap (million €)*
Turkey	85	20	6	1.2	150.6	19,015	–131.6
Spain	46.7	73	7.1	5.2	0.871	1,196.6	43.7
Poland	37.9	60	5	3	49.4	111.6	62.2
Germany	82.8	97	18.1	17.6	878.9	1,272.2	393.3
Italy	60	78	12.3	9.5	386.2	447.1	60.8
Britain	65.8	65	11.4	7.4	135.7	3,166.1	3030.3
France	67	75	12.5	9.4	165.9	516.9	393.3

Source: Tüdam, 2020, accessed on April 2022 from http://www.tudam.org.tr/en/Default.aspx

the importance of using alternative irrigation methods such as using seawater and recycling wastewater especially in rural areas. They also suggest imposing a tropical carbon tax, which is a charge on fossil fuels, to fund natural climate solution projects. Gollakota and others (2022), Kerimray and others (2020), and Liu and others (2020) all agree on the benefits of the pandemic in enhancing the air quality by reducing the energy consumption. They claim that during COVID-19, countries had a chance to prepare the procedures to reduce energy consumption and invest in renewable resources. As a result, after COVID-19, these countries have reduced their coal, oil, and nuclear power in energy consumption.

During curfew, consumers and workers started to use more computers, mobile technology, and the internet to do their daily chores. As a result, the electricity consumption has increased (GU et al., 2020; Ghiani et al., 2020). On the other hand, during the pandemic, many factories stopped their operations and social activities have seized. As a result, electricity consumption decreased significantly (Fezzi and Fanghella, 2020). Given these two opposite effects on electricity consumption, many countries were able to avoid major electricity shortages because of their appropriate electricity consumption regulations (Barbier, et al., 2020; Sobieralski and Hubbard, 2020).

During post pandemic period, Turkey head to new researches to transform linear business models to sustainable models to enable circular economy (CE) in the renewable energy industry through block chains and Erol and others (2021) conducted the first study that investigated solar photovoltaic energy ecosystem (SPVEE) that is based on block chain to increase CE in Turkey. An essential part of sustainable economy is recycling and green insurance. Teufel and others (2019) and Erol and others (2021) illustrated that circular economy can be accessed through block chains in SPVEE to enhance recycling. Therefore, critical success factors should be investigated by the support of the government and entrepreneurs to apply the SPVEE block chains (Makhdoom et al., 2019; Ranta et al., 2021).

Özuyar (2021) claimed that the Organized Industrial Zones, Free Zones, Techno Parks, Clusters, industrial symbiosis are some of the circular economy tools that can affect the water and soil quality towards the economic development and sustainable climate change mitigation. Mukanjari and Sterner (2020) analyzed STOXX Europe 600 index to examine the negative effect of high CO2 emissions production on firm's daily stock returns in energy, air transport, and coke sectors. They claimed that the substitution of basic assets is not the solution, but the asset targeting is the best way for economic growth. Another result suggested that climate change policy is not the way to enhance economic profits in the period of the pandemic. Lahcen and others (2020) assessed the impact of COVID-19 crisis in a small open economy (Belgium) by using a Computable General Equilibrium model. They claimed that the pandemic worsens economies more than the reduction in CO2 emissions. Moreover, they investigated that eco-friendly construction projects and regulations can reduce energy consumption as well as greenhouse gas emissions. Bao and Zhang (2020) used daily air pollution and Intracity Migration Index (IMI) data for forty-four cities in northern China and discovered that air pollution reduction was highly related to travel limitations during the pandemic, and the air quality index (AQI) dropped by 7.80 percent. In addition, their findings emphasize the need of comprehending the role of green production and consumption.

Balbay and others (2021) illustrated that there is a necessity for a long term strategy that concludes waste management and zero waste applications under the concept of "better and more effectively" resource consumption instead of "less." In this way, Turkey can switch its current trade deficit to surplus. More than 80 percent of End-of-Life (EOL) ships are demolished in South Asian countries, including Bangladesh, India, and Pakistan. Except for specific medical-related urgencies, an international supply chain has been broken to a record low due to efforts made to reduce the spread of COVID-19 epidemic (Rahman et al., 2020). They predicted that approximately, 300 million Gross Tonnage (GT), will be accessible for destruction in the next five years, and that the failure to recycle them would cost approximately $20 billion. Another consequence would be economic losses and lost job possibilities in South Asian recycling nations. Rahman and others (2020) contended that a formalized global scale, in conjunction with favorable legislation, may prevent supply chain disruption and boost sustainable development in recipient countries. Bai and others (2021) created a strategy to assist post COVID-19 digital transformation for Micro and Small Enterprises' long-term development. They claimed that digital payments, particularly mobile money, should be a top focus for these firms in their digital transformation. Institutions must also support their resources and competencies in order to implement digital transformation for business continuity and sustainable production and consumption. For sustainability in food sector, Owida and others (2022) suggested that the application of tactical and operational decisions in the production system would help in maintaining sustainability and food safety. Moreover, they shed a light on the technology and efforts of digitalization in order to ensure a sustainable production system during the pandemic.

Health-care systems in high-income countries are increasingly reliant on linear supply chains made up of single-use disposable medical gadgets. This has resulted in higher health-care costs as well as waste and pollution caused by health-care facilities, accompanying with public-health consequences. The conversion of the medical device sector to a more circular economy would help to achieve the objective of delivering increasingly sophisticated treatment in a low-emissions environment (MacNeill, et al., 2020). Corrêa and Corrêa (2021) mention that there are positive effects for COVID-19 pandemic, as the transformation to recycling, remanufacturing, and reusing of raw materials in production. For instance, in Brazil, companies reused raw materials in order to produce masks for involuntary use by citizens. They claimed that the pandemic affected the businesses of sewing experts, with the goal of generating profit and maintaining sustainability. Moreover, the findings showed that the obstacle to circularity is the high demand for masks during the pandemic period.

Circular Economy Organizations of Turkey

Several projects both registered and unregistered have unintentionally contributed to the circular economy model in Turkey. The Ministry of Environment and Urbanization reported that there are twelve groups which have been licensed by the ministry. In partnership with the Ministry of Environment and Urbanization, First Lady Emine Erdoğan launched a statewide Zero Waste Project in 2017. The Ministry of Environment and Urbanization created the Zero-Waste Project in 2017 to avoid resources from ending up in landfills and to raise community awareness about waste management throughout Turkey's transition to CE (Gungor, 2019).

Prior to 2017, the most significant contributors to recycling were unregistered laborers (street garbage collectors) who collected and sorted home waste for a living. They accounted for the majority of Turkey's 9.2 percent garbage recycling systems. The Turkish Government then unveiled its "Zero Waste" program in 2018 in order to enhance garbage management and raise public awareness about waste treatment. All municipalities and public-private partnerships are mandated under the Zero Waste Policy to achieve a specified rate of recycling and reusing per region. The Policy's goal is to raise the domestic solid waste recycling rate to 35 percent by 2023, up from 11.5 percent in 2019. Waste separation and treatment have seen some advances. For instance, the İBB has begun a scheme using reverse vending machines, which allow residents to trade plastic bottles and cans for public transportation credits. Only nine of these vending machines are currently in use in metro stations and schools. This trash management public awareness campaign promotes the usefulness of leftover materials as well. In order to minimize the use of plastic bags, the government imposed a fee on them. This effort boosted the plastic trash awareness campaign while lowering plastic bag usage by 75 percent.

Turkey has been implementing circular economy since 2019 under the banner of sustainability. Due to the effects of the COVID-19 pandemic, climate change, and droughts, this model has been on the agenda since the beginning of 2021, as specified by the EU's Green Deal (Balbay, et al., 2021). In Turkey, there are two organizations that encourage the quick advancement of circular economy studies: Circular Economy Platform of Turkey (Circular Economy Platform of Turkey, 2021) and DCUBE Circular Economy Cooperative (Dcube, 2021). Turkey Circular Economy Platform offers training, funding possibilities, and consultancy services to organizations that are serious about accelerating their transition to a green economy. Sustainable Industry and Circular Economy group, one of the subgroups of the World and The Sustainable Development Association of Turkey (SKD Turkey), seeks to address resource demands, create measuring tools, and provide technical grants. DCUBE Circular Economy Cooperative works to ensure that efficient and sustainable development policies are implemented throughout the world, particularly in Turkey, as a result of the implementation of the Circular Economy Model, particularly in the fields of "agriculture, food, energy, textile, and water." Most Turkish enterprises exporting to Europe have begun to prefer to obtain raw materials and commodities from suppliers who adhere to the Circular Economy.

ÇEVKO is another registered organization that is liable for the extended producer responsibility of around 1,900 Turkish firms (ÇEVKO, 2018). After taking over the industry's recovery obligations, the ÇEVKO Foundation conducts its activities on a contractual basis with various economic operators. These operators are domestic and foreign corporations active in Turkey as fillers, packaging producers, packing recyclers with private labels, large-scale shopping malls and chain stores from food consumption, medicine, chemistry and oil sectors which used glass, metal, plastic, paper, and composite packaging (KPMG, 2022).

In the face of the COVID-19 epidemic, the World Bank reaffirmed its commitment to support Turkey's industrial transformation to a high-technology nation, through three programs: First, the Organized Industrial Zones (OIZ) initiative, which aims to assist in becoming more efficient, ecologically sustainable, and competitive. Second, Turkey Resilient Landscape Integration Project (TULIP), which focuses on improving green infrastructures in the Bolaman River Basin with the goal of developing sustainable infrastructures and ensuring sustainable livelihoods. Third, Turkey's Seismic Resilience and Energy Efficiency in Public Buildings Project, a loan program issued to restore public infrastructure—such as schools, hospitals, government buildings—for better earthquake protection and improved energy efficiency.

Furthermore, the Business Council for Sustainable Development (BCSD Turkey) was established under the leadership of 13 private sector entities, and the Circular Economy Platform Turkey Materials Marketplace was launched, which is an open, digital marketplace aimed at sharing raw materials among industries (BCSD Turkey, 2020).

On the other hand, several Turkey's largest industries in the private sector have integrated the circular economy application system. For instance, one of the textile firms of Turkey Akkim, which is the largest chemical enterprise, focuses on sustainable development research. Akkim aspires to be a model firm in the field of sustainability management in both Turkey and the industry in which it works.

Municipalities have a disproportionate amount of responsibility for waste management since they do not always have the technical or financial resources to collect, filter, and distribute garbage to proper channels. For example, Environment Cleaning Taxes plan to charge polluters a fee depending on facility type and water usage; however, past rates were deemed insufficient to fund all activities. These tariffs were raised by 9.11 percent in 2021 in the aim of relieving some of the load on local governments.

Recent Developments in the Field of Circular Economy in Turkey Post COVID-19 Pandemic

The COVID-19 pandemic is both a health and a socioeconomic crisis, with the potential to exacerbate poverty and inequalities while also jeopardizing global progress towards the Sustainable Development Goals. Turkey confronts severe environmental problems in climate change mitigation and adaptation, clean energy generation, and natural resource preservation and management, all of which inhibit progress towards the Sustainable Development Goals. The COVID-19 reaction and recovery process must strive to avoid reverting to business activities, which would result in increased emissions and other environmental externalities. Greening the recovery effort, on the other hand, could be a means to creating jobs, fighting environmental health hazards, and enhancing future crisis resilience. Investing in renewable energy and improving energy efficiency in industry and buildings has a lot of promise for lowering greenhouse gas emissions, lowering home electricity costs, encouraging the construction industry, and creating jobs.

Turkey tries to encourage sustainable production and consumption through effective resource management and circular economy and the COVID-19 situation has increased the opportunities. It has had opposing effects on lifestyle behavior and the environment. While COVID-19 reduced travel to work and meetings it has resulted in a higher reliance on private transportation rather than public transportation. Furthermore, the rise in online purchases and meal deliveries has led not only to more packaging being discarded, but also to more fossil fuels being burnt to transport things. When compared to the pre pandemic period, online shopping increased by 171 percent in the first months of the pandemic (UN Turkey, 2020).

The amount of garbage, particularly medical waste, has increased. A lot of the personal protection equipment used by health-care workers can only be worn once before being discarded. In addition, waste management has been adversely affected by COVID-19. Increased usage of single-use plastics and other single-use goods as protective face masks in professional, public, and private settings will continue to provide a significant waste management burden, with environmental repercussions such as increased sea pollution. However, due to the COVID-19 outbreak, several municipal recycling programs have been suspended, and the Turkish government has prohibited afflicted residents from sorting their waste.

On the other hand, the demand for energy in industry and transportation has decreased due to COVID-19 crisis, lowering carbon emissions. When compared to 2019, global oil demand fell by a remarkable 9.3 million barrels per day in 2020 (IEA, 2020). The Eleventh Development Plan, the National Energy Efficiency Action Plan, and the National Renewable Energy Action Plan for Turkey all explicitly state Turkey's clean energy goals. Carbon taxes and fossil fuel subsidy reform were employed as green fiscal policy instruments. Turkey prioritizes areas that provide early potential for job creation and economic activity, as well as the development of more efficient and robust energy systems and the reduction of emissions. This includes an emphasis on buildings and transportation, which would simultaneously boost renewable energy efficiency. Turkey has a huge solar energy potential, with 287 GW (71 percent of total installed capacity) in the electricity by 2050 (Erat, S., Telli, A., 2020). The following are some of the Turkish government ministries and other association's practices towards sustainability during post COVID-19 pandemic.

COVID-19 Waste Materials

It is vital to guarantee the effective management of masks, gloves, and other personal hygiene material wastes used in everyday life, in accordance with the steps implemented across the country in the battle against the COVID-19 outbreak that Turkey is experiencing. In the COVID-19 hygiene material waste circular context that was published by the General Directorate of Environmental Management (CYGM) in December 2020, wastes created in quarantine departments of health institutions designated as quarantine units, as well as wastes generated in infirmaries and other medical units of institutions, organizations, and businesses, are managed as "medical waste" and are not combined with other wastes.

On the other hand, masks, gloves, and other personal hygiene material wastes should be collected separately from other wastes, placed in two tightly sealed bags, "household waste and domestic waste," stored for at least seventy-two hours in temporary storage areas, and then delivered to the municipality to be managed in the "other waste" segment (CYGM, 2020).

Medical Supply Materials' Recycling

Single-use medical equipment is one method Turkey has used to achieve circular economy. Tubing, syringes, and masks are all recycled to their original state. Furthermore, long-lasting medical gadgets are another way for Turkey to conserve the environment while lowering health-care costs and pollution. The fundamental cause is the belief that single-use disposables are safer than reusable devices. Single-use disposables are a very easy method of reducing the danger of human error during the reprocessing of reusable devices. Another step towards a circular economy for Turkey is the safe reprocessing and possible resale of discarded medical equipment, as well as ensuring regulatory compliance. Turkey has also provided hospitals with a method for collecting recyclable things like packaging that cannot be reused or reprocessed in order to decrease waste, emissions, and clinical plastics. The commitment to providing high-value care is the first step in achieving a circular economy. Turkey has the potential to enhance facility operations in terms of energy and waste management, as well as encourage doctors to be more resource conscious and adopt environmentally preferable treatments and technology where possibilities are available, through this broader approach. Following such high-value procurement guidelines would help to create more circular and moral supply chains.

Circular Medicine

Pharmaceutical waste management of leftover drugs and the reusing of unused prescriptions that have recently been returned to pharmacies are another type of circularity. In this way the unexpired and unused drugs can be supplied to other patients. The concept of "reuse of medical waste" is a method for minimizing medical waste. Medical waste is defined by the World Health Organization as "expired, unused, spilled, and contaminated pharmaceutical items, medications, vaccines, and serums" (Zeydan, 2019).

Expired medications are destroyed in pharmacies. There is an annual average financial loss of 3397 TL due to the boxes of medicine that were forgotten; the mean number of drugs not used at home is 10.1±7.76 boxes (561 TL); average boxes that are expired 6.2±6.11 boxes (198±315.6 TL). Furthermore, 70 percent of the pharmacists used expired drugs while 68 percent of the pharmacists accepted donations or grant drugs. However, 83 percent of drug donations were made at least once. Medicines destroyed by pharmacies due to expiration were as follows, antihypertensive, antibiotic, analgesic, antidepressant-antipsychotic, and diabetes drugs, respectively (Zeydan, 2019).

Medical waste is collected on a regular basis from around 10,500 different locations inside Istanbul boundaries. Istac gathers around 28,000 tons of medical waste each year using specialized trucks. Collection is accomplished by using cutting-edge technology and protective clothing; vehicles are cleansed and sterilized on a regular basis after each job. The largest medical waste sterilization center in Turkey has been created, with a capacity of 40,000 tons per year (Istac, 2022).

Circular Hospitals

Zero waste strategies for hospitals were implemented after COVID-19. Waste separation boxes were put at strategic locations across hospitals in all units as part of the initiative. On site, hospital trash was segregated and recycled. Furthermore, food waste from cafeterias was donated to animal sanctuaries. For example, Daday District State Hospital" completed a project to store medical waste bins for the protection of personnel and patients. Metals were used to create this project, which is permanently appropriate for emergency service and laboratory as a way of a storage cabinet. It may be used in any health facility to ensure the safety of patients and employees. Furthermore, another effective and high-quality hospital practice is the battery-powered medical waste collecting trolley project, which disposes medical waste instantly to prevent diseases such as COVID-19 from spreading through medical waste and to protect personnel health. This initiative may be implemented in any unit, and medical waste is delivered to storage as soon as possible to decrease the danger of contamination and sickness.

In addition, hospitals aim to decrease paper, plastic, and battery waste by purchasing moneyboxes and storing them temporarily. This zero waste management project was implemented in Turkey in 2021 to stimulate trash evaluation in the direction of a more sustainable economy. The goal of zero waste and zero loss, on the other hand, is to make effective use of scarce natural resources. "Mehmet Akif Inan Training and Research Hospital" is one of these initiatives. The initiative focused on trash storage areas in health-care institutions. Not only this effort is eliminating waste, but it is also reducing environmental dangers. About ten tons of plastic, five tons of paper, and 100 kilograms of battery trash were collected and recycled as a consequence of this operation.

Zero waste bins for hospitals to collect garbage were acquired in December 2020 and temporary zero waste storage sites were built. Following the completion of the installations, the Provincial Directorate of Environment and Urbanization evaluated the compliance of the zero waste management system established in the hospitals with the Zero Waste Regulation, and a "Basic Level Zero Waste Certificate" was given to all public hospitals in the province. It may be used in a variety of public and private institutions, organizations, and public spaces. Preventing waste and waste generation through zero waste applications, ensuring waste recycling by separating wastes at the source, increasing employee safety, satisfaction, and performance through a clean working environment, patient safety and satisfaction, and environmental protection awareness were all created. Due to its visual environmental awareness in the hospital and commitment to recycling, the zero waste application makes a significant contribution to hospital efficiency. Hospitals that are clean and ecologically concerned will be the forerunners of a clean society.

Other circular economy application is the Mersin Provincial Health Directorate, who creates Zero Waste Collection Stand project to protect environment and human health while also contributing to the country's economy via recycling and social responsibility measures. To collect garbage, recycling collection boxes were installed in the textile and book collecting sections. Hazardous garbage (waste vegetable oils, batteries, electronic equipment parts) was collected in a sixty-liter sealed barrel of expired drugs. These hazardous wastes, which are generated at home and work, contribute to recycling when properly collected, but they endanger the environment and human health when discarded carelessly. According to data from the Ministry of Environment and Urbanization's Zero Waste Information System, this application prevented the emission of 454.10 kg of greenhouse gas into the atmosphere, saved 53.45 cubic meters of water, prevented the filling of the 10.42 cubic meter waste landfill area, and saved 16,263 kWh of energy. Furthermore, collecting the books and clothes that are not used at home and donating them to the less fortunate is an example of social duty.

During the COVID-19 epidemic, a special effort was carried out by Hüyük State Hospital in April 2020 to stimulate personnel and decrease stress through gardening. Furthermore, the growing fruits and vegetables will help to reduce the hospital kitchen's costs (Public Hospitals Good Practice Examples, 2021).

Zero Waste Programs

The near zero waste program responds to Turkey's difficulties by aiding private enterprises in implementing cutting edge waste reduction strategies and technologies with high reliability potential, particularly when apparent market failures are limiting acceptance. Three interconnected components were used to create this application. First, $200 million funding includes $20 million in concessional funds from the Climate Investment Funds and Clean Technology Funds to assist qualifying waste reduction and resource efficiency projects suggested by Turkish firms. Secondly, it includes policy conversation with government officials in order to improve legislation and regulatory frameworks and execution. This involves assisting the Ministry of Environment and Urbanization, local governments, and industry groups in building creative and comprehensive frameworks for scaling up waste reduction programs throughout Turkey. Third, it involves raising awareness on resource efficiency policy and financing options to develop a waste minimization knowledge network for government agencies, municipal entities, sector associations, and the private sector. Furthermore, this program tries to disseminate information and lessons learned regarding regulations and best practices that are just a few of the coordination, knowledge sharing, and visibility activities (CIF, 2020).

On the other hand, Turkey materials marketplace was envisioned by this program as a way to bring more businesses together in terms of information sharing and material exchange. This concept was inspired by the US Business Council's comparable marketplace. The European Bank for Reconstruction and Development held consultations with various industrial associations, including the Turkish Industry and Business Association, which represents 85 percent of Turkey's total foreign trade volume excluding energy and 50 percent of total private sector employment, to test its feasibility in Turkey. This outreach aided in the adaptation and refinement of the marketplace model for the Turkish market, highlighting the importance of putting equal focus on information and material exchange components.

Furthermore, under Emine Erdogan's leadership, zero waste projects were launched in 81 provinces and 951 districts chosen by the Ministry of Environment and Urbanization of Turkey. Under this project, waste will be collected and separated according to its type. These wastes will then decompose into organic waste, which will be recycled into new products. Profits in economics will be realized in this manner. This device is known as a top-up automated packaging machine, and it reduces exhaust emissions by using electric motors. Furthermore, these devices will gather packages made of plastic, metal, and glass. Another possibility is that these vending devices will be connected to persons who deliver the packaging trash via phones, speeches, and the internet. By implementing this initiative, the ministry will encourage and benefit from recycling, as well as improve Turkey's circular economy. This project will be implemented in four steps: First, a zero waste team will monitor the project; second, a dual collection system will be implemented, one for glass waste and another for mobile waste; third, waste will be collected from various locations and sent to recycling facilities; and fourth, waste will be pretreated and stored on a regular basis. To achieve zero waste in shopping centers, the program provides equipment in the appropriate locations as well as a temporary storage area for waste to be sent to license recycling facilities. For hospitals, the program, on the other hand, plans and monitors the amount of waste that will be sent to facilities (Zero Waste, 2022).

Circular Energy

Solar photovoltaic (PV) is one of the most promising renewable energy technologies, and it is predicted to be a major renewable energy source for decades to come (Irena, 2019; Chowdhury et al., 2020). The solar PV sector is quickly evolving as a result of developments taking place across the whole supply chain, including solar panel makers and solar energy providers. As a developing country, Turkey should take the required steps and achieve progress toward CE in the solar photovoltaic ecosystem (SPVEE). To overcome the negative environmental effect of conventional linear economic models, the solar PV sector demands more innovation. Critical success factors (CSFs) must be identified and evaluated to ensure positive outcomes. Turkey has also suggested increasing its solar energy–based power generation to 10,000 MW by 2030, based on its Intended Nationally Determined Contributions (INDC, 2020). In March 2021, a tender for 1000 MW of solar photovoltaic (PV) was held. During Turkey's shift to renewable energy, manufacturing and installation of solar PV technologies are likely to develop substantially. PV modules are not yet legally recognized as e-waste in Turkey, despite the fact that the Waste Electrical and Electronic Equipment Directive (WEEE Directive) was approved in 2015. As a result, Turkey's current system is still insufficient to meet the WEEE Regulation's end-of-life (EOL) PV module collection limit (Erat and Telli, 2020).

Circular Agriculture and Water

In 2013, the Food and Agriculture Organization (FAO) projected that food loss from agriculture to consumer in Turkey was as high as 48 percent. To minimize this proportion, the Turkish Industry and Business Association TÜSIAD developed a plan in 2018 to prevent food waste, and food transportation regulations in Turkey have grown harsher as of January 1, 2019 (FAO, 2013).

Turkey's municipalities may achieve a green recovery from COVID-19 pandemic by focusing on ecosystem management and employment intensive projects, which deliver public goods and expand green urban farming. For example, circular agriculture can be achieved by using permanent grass, reprocessing creative plant material into bioenergy, and using agriculture systems to promote sustainability and reduce water streams, as well as regenerative agricultural methods. Turkey, on the other hand, plans to reduce plastic waste by compressing plant material, as wood and bamboo, to be used in recyclable packaging to substitute plastics.

PepsiCo, which generates energy by converting organic wastes in its biogas facility, has successfully cooperated with Aromsa to source alternative organic trash using the TMM platform. PepsiCo used twenty tons of food waste generated by Aromsa in the creation of electricity in a prior deal. According to the findings of the circular voucher research, Aromsa's additional thirty-two tons of garbage was transformed to biogas, and PepsiCo has committed to consistently processing Aromsa's waste, achieving the best energy production efficiency. According to the results of the study, optimal formulation improves the biological performance of Aromsa's organic wastes in PepsiCo's operations, allowing for the production of more electrical and thermal energy (TMM, 2020).

On a daily basis, ninety-five million pieces of bread are consumed in Turkey, yet six million pieces are discarded. The monetary worth of Turkey's 2.1 billion slices of bread wasted per year is estimated to be 1.5 billion TL. Turkey's yearly economic loss as a result of bread waste is equal to the country's profits from wheat exports, where Turkey ranks first in the world. Plastic Move organization, one of the finalists in TÜSAD's "Bu Gençlikte İş Var 2020" start-up competition, proposes a circular economy-based solution to this problem, reintroducing this waste stream into the economy for future usage in the plastic sector. The Istanbul-based startup was inspired by technologies that use corn starch to manufacture bio plastics, but that approach uses resources that might be used to address hunger, despite the fact that 690 million people go hungry every day. Instead, Plastic Move's strategy addresses many SDGs by recycling starch-based agricultural and food waste to create biopolymers without relying on the farming industry (TÜSİAD, 2020).

Furthermore, "Fazla Gıda" is a UN Development Program that focuses on food waste recovery. The food industry's waste provides chances, and they produce economic value with their trash, according to "Waste Utilization Methods in the Food Industry," that nearly one-third of the food produced worldwide is wasted, resulting in a financial loss of $750 billion.

In addition, olive mill wastewater (OMW) is a substantial product of food sector in olive oil producing countries, according to Turkey Materials Marketplace TMM, with a severe environmental effect if not properly handled. One of the projects conducted by the Ministry of Environment and Urbanization of Turkey is the olive sector waste management project to solve the black water problem and prevent the wastewater pollution (CSB, 2021). Another project that is funded by the European Union's Horizon 2020 program, proposes an innovative circular bio economy approach that involves treating olive mill wastewater with non-conventional yeast to produce valuable products like bio-phenols and carotenoids that can be used in a variety of industries.

In Turkey, irrigation accounts for 74 percent of yearly freshwater use, while bottled watered tap water account for 13 percent (Republic of Turkey, Ministry of Development, 2018). Water resource management is a branch of work that connects the economic, environmental, and social components of sustainability. It entails guaranteeing long-term water and sanitation availability, pollution control, and resistance to threats such as scarcity and flooding. Turkey is not a highly water-stressed country, but regional water shortages currently exist, and the situation is likely to worsen in the near future, given the trend towards higher demand. The use of this water in the irrigation of raw vegetables raises the danger of infection with COVID-19. Turkey protects natural resources during post pandemic through the protection of marine areas and their species, assuring of water and air quality through resources management and friendly agriculture system, and assuring of local communities' safety that depend on natural resources in their consumption.

During the epidemic, in addition to the measures taken at the Wastewater Treatment Plants to protect people's health and prevent the spread of the COVID-19 virus; disinfecting treated wastewater used in plant and green field irrigation; not allowing activities such as water activities or grain washing in the stream region where treated wastewater is discharged. Legislatives have been ordered in the "Circular of Measures Regarding the COVID-19 Outbreak and Wastewater Management" to show the appropriate sensitivity in adhering with occupational health and safety procedures and hygiene laws (CYGM, 2020).

Contaminated Materials Recycling

A comprehensive application was submitted to the Turkey Circular Economy Platform, which aided member firms in their exploration of circular economy potential and the creation of both environmental and economic consequences. Bilecik Demir Çelik took advantage of the chance to conduct research into the possibilities of producing secondary raw materials from refractory waste that is currently being utilized in the steel plant. As the first stage in the project, the quantity of trash created by the firm was cut in half by the process management. After that, new methods for repurposing these wastes as refractory materials in the company's operations were successfully developed. As a consequence of the study, the firm not only saved €550,000, but also avoided the use of 1,200 tons of bauxite, 1.2 million kwh of electricity, and 65,000 sm3 of natural gas. A total of 1,400 tons of red mud and 1,200 tons of CO2 emissions were avoided as a result of the project. Furthermore, by lowering the usage of "White Fused Alumina (WFA)," which is entirely imported, the firm was able to realize significant environmental and economic benefits (TMM, 2020).

The deal between Arkim and Nurol Construction is another example of circular economy. Arkim obtained secondhand steel construction parts from Nurol, which had shuttered one of its facilities, for use in its new production plant. Nurol approached the TMM team to find a new home for their used materials on the platform, and successfully connected with Arkim, which was building a new facility. Arkim, which is developing its capacity in generating natural food preservatives from eggshells, teamed with Nurol Construction using the TMM platform.

Moreover, companies that offer products to the market in reusable packages will prepare and submit a deposit plan through the Packaging Information System of the Ministry of the Environment and Urbanization, as required by the regulations. If the Ministry deems it necessary, this plan will be given a code number and carried out in accordance with the infrastructure program that was in practice in 2020.

Circular Plastic

In Turkey, there are around 1,450 plastic packaging and 5,000 plastic processing companies. The plastic production exceeds 4.1 billion tons and plastic packaging production is about nine million tons. Moreover, total plastic industry exports exceeded five billion in 2020 and offered jobs for more than 160,000 employees (Republic of Turkey Ministry of Trade, 2021). Toward circular economy prospective, the Turkish government posed a fee on the used plastic bags, which went into effect on January 1, 2019. This has already decreased the use of plastic bags by 80 percent, and the adoption of deposit bottles is also planned for 2021–2023 (Daily Sabah, 2019).

During the COVID-19 lockdown there was an increase in plastic consumption. In 2020, the output of plastic raw materials was estimated to be over one million tons, accounting for 42 percent of the total production of the Turkish packaging sector (Advanced Circular Packaging, 2020). Turkey got the most plastic garbage from Europe in 2020, receiving a total of 659,960 tons from the EU and the United Kingdom (NTV, 2020). In addition, Turkey is Europe's top waste importer, getting 13.7 million tons out of a total of 32.7 million tons of rubbish from the EU each year (Eurostat, 2020).

Concerns about hygiene arise in the food sector when recycled materials are used. Plastic packaging is frequently used for food storage and plays an important role in nutrient preservation. Despite the fact that they are not environmentally friendly materials, their role in the food sector is crucial for decreasing food waste. Because they come into direct contact with edible items, they must adhere to certain health standards, which are outlined in the Turkish Food Codex Legislation. Understandably, health standards slow the development of sustainable substitute materials that serve the same aim as plastic recycling.

The reuse material circularity of Turkey was calculated to be 7 percent. Increasing this performance will require the implementation of new regulations that encourage the use of circular materials during manufacturing, such as establishing a minimum percentage of recycled and renewable content or providing tax breaks to companies that achieve a certain level of circularity input. However, the output circularity was calculated to be 13 percent (Pagev, 2022). PAGEV's business members account for 88 percent of Turkey's plastics income. "Operation Clean Sweep," which focuses on minimizing micro plastics from manufacturers, is one example of a project performed by PAGEV members to assist combat plastic pollution (Operation Clean Sweep, 2021).

Food wastes converted into a bio-plastics input stream is another practice towards circularity. For instance, Wasteespresso is a Turkish entrepreneur that gathers, composts, and dries coffee waste from restaurants, hotels, and cafes in order to manufacture raw bio-plastic materials. Furthermore, the Damla 330 milliliter plant bottle by Coca-Cola is made of bio-based materials. In 2019, Aghita Group PSJC created the first 100 percent recyclable plant water bottle (Wastespresso, 2022).

The demand for plastic increased during COVID-19. There has been an increase in demand for single-use plastics such as surgical masks, disinfection containers, and utensils in Turkey and throughout the world. As more ecologically friendly items are demanded by Turkish customers, producers have begun to use bamboo or wood as a plastic replacement (PAGEV, 2021).

Zero CO_2 Emissions and Sustainable Environment

For more sustainable environment, October 6, 2022, represented a significant milestone in Turkey's international and national sustainability policy, with the Parliament finally ratifying the Paris Agreement, more than five years after signing the pact. Despite being a member of the OECD, Turkey has long claimed that countries should be fined based on their past emissions. The agreement's long-term aim is to keep global temperature rise below 2°C, and to restrict it to 1.5°C if possible. As a result, all parties have made it a priority to emphasize the use of renewable energy sources while reducing the usage of fossil fuels. Under the Paris Agreement, Turkey has pledged to reduce its emissions by 21 percent by 2030 and become net-zero by 2053. The Climate Change Strategy and National Contribution Statement for Turkey will be revised accordingly. Some environmental initiatives are already underway or have been completed. For example, the Green Deal Action Plan 2021, which is the first significant publication recognizing the relevance of circular concepts and harmonization with the European Green Deal was released in July 2021. The infrastructure for the Emissions Trading System is being created to facilitate investments in climate friendly and clean industrial technologies.

Another activity for Presidency's Annual Program is the Environment and Climate Action Sector Operational Program, "An Overall Assessment of Turkey's Potential in the Context of Transition to Circular Economy Project," was launched on February 22, 2022, under the Environmental Management for Sustainable Development. This was co-funded by the European Union and the Republic of Turkey, with the Ministry of Environment, Urbanization and Climate Change designated as the ultimate beneficiary. The project budget will last thirty-six months, which is roughly 3.2 million Euros, with the EU contributing 85 percent and the Turkish government contributing 15 percent. The project intends to strengthen Turkey's institutional and technical capabilities in conformity with the EU Circular Economy Package, particularly in waste management and the transition to a circular economy (Climate Home News, 2022).

One of the important practices to reduce CO_2 emissions is to substitute glass bottles for plastic bottles. Coca-Cola (CCI) launched the returnable glass bottles project in both Turkey and Pakistan in 2020 to extend the life cycle of their products and lower the company's energy and resource usage in bottle manufacturing. These bottles may be reused up to eight times, reducing CO_2 emissions by 67,590 tons, which is equivalent to the CO_2 intake of 5.6 million mature trees. The biopolymer was created in collaboration with firms in the Fast-moving consumer goods (FMCG) and automotive industries, both of which have significant levels of plastic use; their packaging partners enable these companies to cut their usage of plastic by up to 20 percent. These businesses want to lower their carbon footprint and meet their environmental goals faster by limiting their use of plastic (Pagev, 2020).

During COVID-19, the Turkish Government posed restrictions on using vehicles for specific ages. The exhaust gas emission measurements of cars belonging to residents aged sixty-five and over, as well as those with chronic diseases who are forbidden due to curfew, have been postponed as of the curfew date, as part of the Ministry of Environment and Urbanization's initiatives to safeguard citizens from the COVID-19 outbreak. After the restriction is lifted, exhaust gas emission measurements can be taken within forty-five days, following the conclusion of that time, additional car owners will be able to get an exhaust emission measurement (CYGM, 2020).

Conclusions and Recommendations to the Turkish Community to Empower Circular Economy

Turkey considers that it is critical to maintain its human centered development model, with the major axes of equality and sustainability. At present, Turkey's top goal is to combat poverty and maintain a respectable quality of life among all citizens. Because accelerating sustainable development necessitates the mobilization of financial resources, it is critical to maximize the use of existing financial resources while also fostering innovative collaboration for future partnerships. Moreover, there is a requirement for additional financial sources, so private sector investments might be used. In this regard, the relevance of the instruments mentioned in this chapter will be stressed since they are good instances of economic development in tandem with a sustainable environment. Nonetheless, it should be mentioned that all of these tools should be examined, and stakeholder assessments should be conducted to extend this effect. Furthermore, these tools should be reassessed for socioeconomic importance.

The COVID-19 epidemic increases the need for public knowledge of waste management through usable models and zero waste applications. In this way, in addition to recycling waste materials, it will be possible to use resources more effectively and efficiently by delivering them as needed and offering them for consumption. It is vital to rebuild the linear chain into a sustainable one and to base the waste management process on recycling. Innovations in product design and material engineering, and computational modeling will be achievable with the specialists who will be taught in this process. In this way, people's behaviors will be redirected towards waste management and anti-pollution actions. In addition, manufactures will preserve the goods and materials utilized for a longer length of time to protect the natural systems.

A dynamic interaction of health-care stakeholders has resulted in the extensive growth of single-use disposables in the health-care supply chains. The costs of linearity, including environmental and public health consequences, must be integrated as part of a commitment to high-value treatment in order to identify and critically analyze the full cost of modern health-care. A circular health-care economy based on resource efficiency, reuse, and recycle provides an alternative to the unsustainable implications of existing practices. This involves a broad industry change prompted by demand, creative revenue models, and legislative and professional regulations that prioritize public health, sustainability, and patient's health protection.

Reference

Abbott A. (2019). "Europe's next €1-billion science projects," 566, 164–65, retrieved on March 3, 2022, from https://www.nature.com/articles/d41586-019-00541-y.

Advanced Circular Packaging (2020). accessed on April 4, 2022 from https://www.advancingcircularpackaging.com/sustainable-plastic-packaging/resins-and-types-of-packaging/.

Andersen, M., et al. (2006). "An introductory note on the environmental economics of the circular economy," Sustainability Science, *2*(1):133–40, retrieved on Feb 3, 2022, from https://doi.org/10.1007/s11625-006-0013-6.

Bai, C., et al. (2021). "COVID-19 pandemic digitization lessons for sustainable development of micro-and small- enterprises," Sustainable Production and Consumption, retrieved on March 3, 2022, from https://doi.org/10.1016/j.spc.2021.04.035.

Balbay, S., et al. (2021). "Circular Economy/Industrial Sustainability" Approach in the World and in Turkey," European Journal of Science and Technology, No. 27, pp. 557–69, retrieved on Feb 3, 2022, from https://doi.org/10.31590/ejosat.971172.

Barbier, E. B. (2020). "Is green rural transformation possible in developing countries?" World Development, 131, 104955, Retrieved on March 3, 2022, from https://www.nature.com/articles/d41586-020-00324-w.

Barbier, E. B., Lozano, R., Rodriguez, C. M., and Troeng, S. (2020). "Adopt a carbon tax to protect tropical countries Nature," 578, 213–16, retrieved on March 3, 2022 from https://doi.org/10.1016/j.worlddev.2020.104955.

Bao, R., Zhang, A. (2020). "Does lockdown reduce air pollution? Evidence from 44 cities in northern China," Science of the Total Environment, retrieved on March 12, 2022, from https://doi.org/10.1016/j.scitotenv.2020.139052

Business Council for Sustainable Development Turkey (2020). "Sustainable Industry and Circular Economy," retrieved on Feb 13, 2022, from http://www.skdturkiye.org/en/surdurulebilir-sanayi-ve-donguselekonomi.

Chowdhury, et al. (2020). "An overview of solar photovoltaic panels' end-of-life material recycling," Energy Strategy Reviews, 27, Article 100431, retrieved on Feb 17, 2022, from https://doi.org/10.1016/j.esr.2019.100431.

Climate Home News (2022). "Turkey ratifies the Paris Agreement after approving a 2053 net zero goal" retrieved on Feb 17, 2022, from: https://www.climatechangenews.com/2021/10/06/turkey-ratifies-paris-agreement-approving-2053-net-zero-goal/.

Climate Investment Fund (2021). "Near Zero Waste in Turkey, Moving toward a Circular Economy by Monetizing Waste" retrieved on Feb 10, 2022, from: https://www.effectivecooperation.org/content/near-zero-waste-turkey-moving-toward-circular-economy-monetizing-waste.

Climate Investment Fund (2020). "Near Zero Waste in Turkey" retrieved on Feb 12, 2022, from: https://www.climateinvestmentfunds.org.

Corrêa, H., Corrêa, D. (2021). "The COVID-19 Pandemic Opportunities for Circular Economy Practices among Sewing Professionals in the City of Curitiba-Brazil," Frontiers in Sustainability, retrieved on March 7, 2022, from: https://doi.org/10.3389/frsus.2021.644309.

CSB. (2021). The official site of the Ministry of Environment and Urbanization of Turkey, retrieved on March 13, 2022, from: https://strateji.csb.gov.tr/.

CYGM. (2020). "Circular of Measures Regarding the COVID-19 Outbreak and Wastewater Management" retrieved on March 13, 2022, from https://cygm.csb.gov.tr/COVID-19-salgini-ve-atiksu-yonetimine-iliskin-onlemler-genelgesi-duyuru-407943.

CYGM. (2020). "The COVID-19 hygiene material waste circular has been published" retrieved on March 13, 2022, from https://cygm.csb.gov.tr/COVID-19-hijyen-malzeme-atiklari-genelgesi-yayimlandi.-duyuru-407923.

CYGM. (2020). "Circulars on Exhaust Gas Emission Measurements during the COVID-19 Outbreak" retrieved on March 3, 2022, from https://cygm.csb.gov.tr/COVID-19-salgini-suresince-egzoz-gazi-emisyon-olcumlerine-iliskin-genelgeler-duyuru-407968.

Circular Economy Platform of Turkey. (2021). official site, retrieved on April 3, 2022, from https://donguselekonomiplatformu.com/.

Erat, S., Telli, A. (2020). "Within the global circular economy: A special case of Turkey towards energy transition," MRS Energy and Sustainability, retrieved on March 3, 2022, from https://doi.org/10.1557/mre.2020.26.

Erol, et al. (2021). "Towards a circular economy: Investigating the critical success factors for a blockchain-based solar photovoltaic energy ecosystem in Turkey." Energy for Sustainable Development, retrieved on Feb 27, 2022, from https://doi.org/10.1016/j.esd.2021.10.004.

Eurostat. (2020). European Union statistic, retrieved on April 6, 2022, from: https://ec.europa.eu/eurostat/web/products-eurostat-news/-/ddn-20210420-1.

Fezzi, C., Fanghella, V. (2020). "Real Time Estimation of the Short-Run Impact of COVID-19 on Economic Activity Using Electricity Market Data." Environmental and Resource Economics (2020) 76:885–900, retrieved on March 4, 2022, from https://doi.org/10.1007/s10640-020-00467-4.

Ghiani, E., Galici, M., Mureddu, M., Pilo, F. (2020). "Impact on Electricity Consumption and Market Pricing of Energy and Ancillary Services during Pandemic of COVID-19 in Italy." Department of Electrical and Electronic Engineering, University of Cagliari, retrieved on March 4, 2022, from https://doi.org/10.3390/en13133357.

Gollakota, A., Chi-MinShu. (2022). "COVID-19 and energy sector: Unique opportunity for switching to clean energy." Gondwana Research, retrieved on March 4, 2022, from https://doi.org/10.1016/j.gr.2022.01.014.

Gu, S., Kelly, B., Xiu, D., "Empirical Asset Pricing via Machine Learning." The Review of Financial Studies, Vol. 33, Issue 5, May 2020, Pages 2223–73, retrieved on March 4, 2022, from https://doi.org/10.1093/rfs/hhaa009.

Gungor, E. (2019). "Cycle or not to cycle: Towards a circular economy in Turkey." Netherlands Enterprise Agency, retrieved on Feb 11, 2022, from https://www.rvo.nl/sites/default/files/2019/10/To-cycleor-not-to-cycle.pdf.

Istac. (2022). "Medical Waste Management," retrieved on April 23, 2022, from https://www.istac.istanbul/en/clean-istanbul/our-services/istanbul-medical-waste-management.

IEA. (2020). Oil Market Report, retrieved on Feb 23, 2022, from https://www.iea.org/reports/oil-market-report-april-2020.

Intended Nationally Determined Contribution. (2020), retrieved on Feb 5, 2022, from https://www4.unfccc.int/sites/submissions/INDC/Published%20Documents/Turkey/1/The_INDC_of_TURKEY_v.15.19.30.pdf.

IRENA. (2019). "Future of solar photovoltaic." retrieved on Feb 5, 2022, from https://www.irena.org/publications/2019/Nov/Future-of-Solar-Photovoltaic.

KBMG. (2022). "Preliminary Research on Circularity Potential of Five Sectors in Turkey." retrieved on March 5, 2022, from http://www.skdturkiye.org/en/news/we-held-the-second-turkey-circular-economy-week.

Kerimray et al. (2020). "Assessing air quality changes in large cities during COVID-19 lockdowns: The impacts of traffic-free urban conditions in Almaty, Kazakhstan." Science of The Total Environment, retrieved on March 2, 2022, from: https://doi.org/10.1016/j.scitotenv.2020.139179.

Kilickaplan, et al., (2017). "An energy transition pathway for Turkey to achieve 100% renewable energy powered electricity, desalination and non-energetic industrial gas demand sectors by 2050." Solar Energy 158, 218–35, retrieved on March 1, 2022, from https://doi.org/10.1016/j.solener.2017.09.030.

Liu et al. (2020). "Near-real-time monitoring of global CO2 emissions reveals the effects of the COVID-19 pandemic." Nature Communications, retrieved on Feb 20, 2022, from https://www.nature.com/articles/s41467-020-18922-7.

Makhdoom et al. (2019). "Blockchain's adoption in IOT: The challenges, and a way forward," Journal of Network and Computer Applications, 251–79, retrieved on Feb 22, 2022, from https://doi.org/10.1016/j.jnca.2018.10.019.

Mukanjari, S., Sterner, T. (2020). "Charting a "Green Path" for Recovery from COVID-19." Environmental and Resource Economics (2020) 76, 825–53, retrieved on March 10, 2022, from https://doi.org/10.1007/s10640-020-00479-0.

NTV. (2020). "Greenpeace: Türkiye, Avrupa'dan 659 bin 960 ton plastik atık ithal etti," retrieved on April 7, 2022, from https://www.ntv.com.tr/galeri/turkiye/greenpeace-turkiye-avrupadan-659-bin-960-ton-plastik-atik-ithal-etti,M_8O5Qq3SE-oHYZMw-ArcQ.

Operation Clean Sweep. (2021), official site, retrieved on April 12, 2022, from: https://www.opcleansweep.org/.

Özuyar, P. (2021). "Circular Economy Practices In Turkey and Their Potential Interaction with Climate Change Mitigation and Adaptation." ARHUSS, (2021), 4(1):18–32, retrieved on May 3, 2022, from https://acikerisim.istinye.edu.tr/xmlui/handle/20.500.12713/1850.

Owida, A., et al. (2022). "Decision-making framework for a resilient sustainable production system during COVID-19: An evidence-based research." Computers and Industrial Engineering, retrieved on May 3, 2022, from https://doi.org/10.1016/j.cie.2021.107905.

Pagev. (2020), retrieved on March 12, 2022, from https://pagev.org/upload/files/T%C3%BCrkiye%20Plastik%20Sekt%C3%B6r%20%C4%B0zleme%20Raporu%202020%20revize%20%283%29.pdf.

Pagev. (2022). "Türkiye Plastik Sektör İzleme Raporu 2020," retrieved on April 17, 2022, from https://pagev.org/.

Pagev. (2021). "Koronaya karşı tek kullanım!" retrieved on April 20, 2022, from https://pagev.org/koronaya-karsi-tek-kullanim.

Public Hospitals Good Practice Examples. (2021), retrieved on March 13, 2022, from https://iyiuygulamaornekleri.saglik.gov.tr/TR,62404/atik-yonetimi-cevre-yonetimi.html.

Ranta, V., et al. (2021). "Digital technologies catalyzing business model innovation for circular economy multiple case study." Resources, Conservation and Recycling, 164, Article 105155, retrieved on Feb 4, 2022, from https://doi.org/10.1016/j.resconrec.2020.105155.

Rahman et al. (2020). "Disruption in Circularity? Impact analysis of COVID-19 on ship recycling using Weibull tonnage estimation and scenario analysis method." Resources, Conservation and Recycling, retrieved on Feb 4, 2022, from: https://doi.org/10.1016/j.resconrec.2020.105139.

Republic of Turkey, Ministry of Development. (2018). "11th National Development Plan of the Republic of Turkey, Water Resources Management and Security Special Expertise Commission Report." Ankara, retrieved on Feb 28, 2022, from https://www.urbanagendaplatform.org.

Republic of Turkey Ministry of Trade. (2021). "Plastic Processing Industry." retrieved on April 28, 2022, from: https://www.trade.gov.tr/data/5b8fd4fb13b8761f041fee78/Plastic%20Processing%20Industry.pdf.

Teufel, B., et al. (2019). "Blockchain energy: Blockchain in future energy systems." Journal of Electronic Science and Technology, retrieved on Feb 6, 2022, from: https://doi.org/10.1016/j.jnlest.2020.100011.

TÜSİAD. (2020). "Plastic Move–Büşra Köksal | TÜSİAD Bu Gençlikte İŞ Var! 2020 Finalisti," retrieved on April 5, 2022, from https://www.bugenclikteisvar.com/index.php/program/bizden-haberler/428-plastic-move-busra-koksal-tusi-ad-bu-genclikte-i-s-var-2020-finalisti.

Turkey Materials Marketplace. (2020). "15th Application: Bilecik Demir Çelik." retrieved on Feb 6, 2022, from: https://ebrd-tmm.com/news/post/15-uygulama-gerceklesti.

UN, Turkey. (2020). "COVID-19 Socioeconomic Impact Assessment Report." retrieved on Feb 14, 2022, from: https://turkey.un.org/en/126064-un-turkey-socio-economic-impact-assessment-report.

UN, Turkey. (2021). "COVID-19 Socioeconomic Response offer," retrieved on Feb 13, 2022, from: https://turkey.un.org/en/126066-un-turkey-COVID-19-socio-economic-response-offer.

UN Sustainable Development Group. (2022). "Universal values. Principle two: Leave no one behind," retrieved on April 30, 2022 from: https://unsdg.un.org/2030-agenda/universal-values/leave-no-one-behind

Wang, Q., and Su, M. (2020). "Integrating block chain technology into the energy sector—From theory of block chain to research and application of energy block chain." Computer Science Review, 37, Article 100275, retrieved on March 10, 2022, from https://doi.org/10.1016/j.cosrev.2020. 100275.

Wang, Q., Li, R., and Zhan, L. (2021). "Block chain technology in the energy sector: From basic research to real world applications." Computer Science Review, 39, retrieved on March 14, 2022, from https://doi.org/10. 1016/j.cosrev.2021.100362.

Wang, Y., Singgih, M., Wang, J., and Rit, M. (2019). "Making sense of block chain technology: How will it transform supply chains?" International Journal of Production Economics, 211, 221–36, retrieved on March 1, 2022, from https://doi.org/10.1016/j.ijpe.2019.02.002.

Wastespresso. (2022), official site, retrieved on April 2, 2022, from: https://www.wastespresso.com/.

World Health Organization. (2021). "Interactive platform visualizing mortality data," retrieved on Feb 15, 2022, from: https://www.who.int/data/mortality/country-profile.

Yadav, S., Singh, P. (2020). "Blockchain critical success factors for sustainable supply chain Resources." Conservation and Recycling, 152, retrieved on March 5, 2022, from: https://doi.org/10.1016/j.resconrec. 2019.104505.

Yildizbasi, A. (2021). "Blockchain and renewable energy: Integration challenges in circular economy era," Renewable Energy, 176, 183–97, retrieved on March 3, 2022, from: https://doi.org/10.1016/j.renene. 2021.05.053.

Zero waste. (2020). official site of the "Zero Waste" project, retrieved on March 13, 2022, from: https://zerowaste.gov.tr/.

Zeydan, V. (2019). "Drugs Returned to Pharmacies Pharmacoeconomic Analysis," Istanbul Medipol University Istanbul, Health Sciences Institute, official site, retrieved on April 13, 2022, from: https://www.medipol.edu.tr/en/.

Conclusion

NATALYA KETENCI

Environmental degradation in the world is one of the most discussed problems in the literature for many decades. There are a lot of factors discussed that worsen our environment. Environmental degradation due to the pollution from fossil fuels induces countries to decrease their use, however, until the productivity of renewable resources reaches the necessary level, countries continue to be highly dependent on non-renewable sources. Another important issue is increasing waste in all spheres of our life. One of the solutions for the environmental degradation slowdown is the circular economy. The circular economy aims to produce as less waste as possible by reusing materials in new productions. The circular economy is becoming part of many different sectors of an economy. This book discusses and analyzes different sectors that are widely becoming to be involved in the circular economy process in Turkey. The research shows that the circular economy is an inevitable part of our life if we want to: save scarce and valuable resources for the future generations; decrease dependence on imported resources; save our environment from degradation; to improve health standards by decreasing the risk of being damaged by the pollution, and there are many other reasons why it is time to take it seriously.

Index

About the Contributors

Zehra Binnur Avunduk received BSc (Honors), MBA, and PhD degrees in business administration from the Faculty of Economics, Istanbul University, Beyazıt, Istanbul, Turkey, in 2009, 2012, and 2017, respectively. She is currently a lecturer in the Vocational School of Social Sciences, Istanbul University-Cerrahpaşa, Istanbul, Turkey. Her research interests include innovation strategies, open innovation, sustainability, entrepreneurship, technology, production, and operations management. Dr. Avunduk has published articles in national and international peer-reviewed journals, and book chapters. She also served as an ad-hoc reviewer in several prestigious journals and presented many papers at scientific conferences. She has received the "Best Paper Award" from the 2018 International EMI Entrepreneurship and Social Sciences Congress. Dr. Avunduk also has served as a reviewer for Academy of Management annual meetings. She has been a researcher in a project funded by the Istanbul Development Agency (ISTKA).

Yusuf Can Erdem has obtained his undergraduate degree in business administration from Middle East Technical University, Ankara, Turkey. After graduation he was honored with a state scholarship to get his MBA degree in the United States. Following his graduation from East Carolina University, North Carolina, with an MBA degree, he has assumed sales and marketing responsibilities at different companies. Following his retirement from professional business life, he has received his PhD in business administration from Yeditepe University in 2007. Ever since he has taught courses of e-business, supply chain management, consumer behavior,

CRM, marketing research, international marketing, marketing strategy in both undergraduate and graduate schools of different universities. Two of his books on Customer Experience Management have been published in Turkey and the United States. He is currently employed in the management department of Yeditepe University.

Alper Altınanahtar received his bachelor's degree in public finance from Ankara University in 1997. Following his bachelor's degree, he worked as a research assistant at Faculty of Economics and Administrative Sciences at Mediterranean University. In his last year at Mediterranean University he went to Texas Tech University, United States, where he received his MA (2001) and PhD (2004) degrees in the field of economics. During his stay at Texas Tech University, he was employed as a teaching assistant, then graduate part time instructor and between 2003 and 2005 he was appointed as an assistant professor. Since 2006 Mr. Altınanahtar has been working at Yeditepe University, Department of Economics. Currently he is an associate professor in economics and appointed as the head of the Business Administration Department and as the vice dean of the Faculty of Economics and Administrative Sciences. His works have been focused on various economics topics such as human organ shortage and procurement, socio-economic determinants of suicides and anti-smoking policies and their effects on smoking prevalence, health care finance, and energy markets.

Tulin Johansson received her PhD in agricultural economics, specializing in environmental economics from the University of Kentucky in 2000, with her dissertation on the interactions between economic development and environmental quality. She received her Master of Science degree in economics also from University of Kentucky and her bachelor's degree in architecture from Istanbul Technical University. Tulin Johansson is the Global Managing Director at Spectrum House, Istanbul. Prior to joining the Spectrum House, Tulin Johansson served as associate professor of economics at Lasell University between 2005 and 2019. She also served as a chair of the accounting/finance/economics department for seven years at Lasell University. In addition, Tulin Johansson held lecturer positions at Northeastern University and the University of Boston, Massachusetts. She has published and presented articles on environmental quality and economic development. Her current scholarly interest lies on the Circular Economy, and she is currently enrolled in "Economics and Policies for a Circular Bio-Economy" MicroMaster program at Vageningen University.

Meral Özalp worked eight years as a lecturer in accounting and tax applications in Akdeniz University between the years 2011–2020. Prior to that she worked two years in operations in marketing. As she was working as a lecturer, she pursued PhD studies in agricultural economics in Akdeniz University and received her PhD degree in March 2019. She holds a master's degree in economics from Marmara University and a bachelor's degree in business administration from Trakya University. Besides that, she worked as a researcher and project consultant for several scientific and European Union projects. She has several articles on agricultural policy.

Ayşe Gönül Demirel is associate professor of management and strategy in the Department of Business Administration at Yeditepe University. She has earned her baccalaureate degree from Bogazici University, Istanbul, majoring in sociology; received a certificate in marketing from the Chartered Institute of Marketing, UK; received her MBA in management and organizations from Yeditepe University, Istanbul; and PhD. in management and organization from Yeditepe University in 2008 with a thesis on "An Exploratory Study on Developing and Testing Integrated Stakeholder Relationships Management Model" under the supervision of Professor Dr. M. Atilla Öner. She has been teaching courses at the Department of Business Administration since 2004 in undergraduate, graduate, and doctorate levels. Her teaching experience includes courses such as: organization theory and design, strategic management, business research methodology and data analysis. human resources management, and corporate governance. Her current research interests are stakeholder theory, managerial values, leadership, and corporate governance. Her recent research focuses on the influence of Big Data on management and organization issues. She has authored and coauthored several articles and international book chapters.

Veysel Karani Nurdağ completed his undergraduate education at Yıldız Technical University as double major student in industrial engineering and mathematics engineering simultaneously and as the top student of the department. He earned his master's degree at Bahçeşehir University, at the Department of Business Administration. He is currently continuing his academic studies at the thesis stage in the doctoral program at Yeditepe University, Department of Business Administration. The title of his thesis is "The mediating role of psychological capital on the effect of agile leadership on Employee Intrapreneurship: A research in Turkey." Other current research interests include sustainability management issues.

Tülin Ural is professor of marketing at the Business Administration Department, Faculty of Economics and Administrative Sciences, Yeditepe University, İstanbul, Turkey. She received a bachelor's degree in economics from Istanbul University. She was granted by Tubitak-Bideb post-doctoral scholarship and went to Michigan State University, United States, The Eli Broad College of Business, as visiting professor, in 2006–2007. She has written books, including *Business and Marketing Ethics* and *Road Map of Branding*. She has been published extensively as author and co-author of over fifty papers in highly regarded, peer-reviewed journals. She focuses on advanced quantitative research techniques, multivariate data analysis, strategic marketing management, marketing ethics, brand management, consumer behaviors, and other marketing issues for organizations.

Oğuz Oypan completed his undergraduate education at Anadolu University, Department of Business Administration in 2012. He also received the Ph.D. degree in Hatay Mustafa Kemal University, Department of Business Administration in 2021. He worked as a research assistant in Hatay Mustafa Kemal University, Marketing Department between 2018 and 2021. He still works as a lecturer in Hatay Mustafa Kemal University, Antakya Vocational School, Real Estate Management Program. He has scientific studies in the fields of consumer behavior, sustainable consumption, and digital marketing.

Ozan Atik was born in Samsun in 1989. He holds a BBA degree from Uludağ University (2012), and an MBA from Yeditepe University (2021). His thesis examined the effect of subjective norms, environmental awareness, environmental concern, and attitude on purchase intention towards green home. He has worked in the heating, ventilation, air conditioning (HVAC) industry as a sales executive. He is currently working as a senior sales specialist at ODE insulation company, which supplies insulation materials to the construction industry. His research interests include green marketing, green buildings, and consumer buying behavior.

Nilgün Gürkaynak was born in Izmir in 1967. She is a graduate of Izmir American Collegiate Institute (1985). She holds a BBA degree from Boğaziçi University (1989), and an MBA from Dalhousie University (1991) and a PhD in Marketing from Yeditepe Unviversity (2008). She served as a faculty member of marketing in Yeditepe University between 2000 and 2009; before that, worked in various managerial positions for the Coca-Cola Company, GfK, Henkel-Turyağ and Izmir Cotton Textile Mills (IPM T.A.Ş). Gürkaynak also provided consultancy and training services for firms from FMCG, banking, construction materials, real estate and communication sectors. She joined IUE in 2009 and between 2010 and 2013 was the director of the School of Applied Managerial Sciences. She took part in projects carried out by the Ministry of Industry and Technology, Ministry of Interior, Ministry of Agriculture and Forestry, Chambers of Commerce and Exporters' Union. Her research interests include place marketing, marketing for SMEs, social media, and local brands.

Muteber Tuzcu is a research assistant at the Business Administration Department, Faculty of Economics and Administrative Sciences, Yeditepe University, İstanbul, Turkey and a master student at Department of Management, Institute of Social Sciences, Boğaziçi University, İstanbul, Turkey. She received a bachelor's degree in tourism administration from Boğaziçi University. Her research interests include services marketing, consumer behaviors and sustainability.

Özge Kirezli received her MA degree in production management and marketing from Marmara University in 2010. In 2015 she received her PhD degree in production management and marketing from Marmara University. She started her academic career as a teaching assistant in the Business Administration Department of İstanbul Bilgi University in 2010, and worked there until 2018. Since 2019, she is working as an assistant professor at Yeditepe University, lecturing on marketing principles, marketing management, marketing research, and consumer behavior courses. She has articles in The Journal of World Intellectual Property and Journal of Innovation Economics; Management and book chapters about sharing economy, green marketing and corporate social responsibility in Global IGI books and Routledge. Her research interests are consumer decision making and paradigm shifts in contemporary marketing.

Kubilay Düzgit is working as a research assistant at the Business Administration Department, Faculty of Economy and Administrative Sciences of Yeditepe University. He received his bachelor's degree in business administration from Istanbul University. He is currently studying accounting and finance at Marmara University. His research area covers green innovation, sustainability, corporate social responsibility, asset valuation, firm financial stability, and firm financial performance.

Mehmet Nuri Önortaç graduated from METU, Department of Economics and Statistics. He finished his postgraduate study at Birmingham

University, in the field of mathematical economics. His research topic was related to the application of linear programming to Turkish agriculture. Presently he is an accounting and finance lecturer at Yeditepe University, Department of Business Administration.

Gül Esin Delipınar graduated from Maltepe University with a PhD in international trade and logistics in 2020. Her research areas are international trade, business, and tourism. In 2021, she lectured at Yeditepe University Transportation Management. She is self-employed and planning to pursue her career in academics.

Ayşe Sevencan Akcabelen is assistant professor in economics at Yeditepe University. Since 2022, she is the assistant director of the Management Application and Research Center at Yeditepe University (MARC-YUVAM). She received her PhD in Economics from Yeditepe University (2010), Master of Arts in applied economics from the University of North Carolina (2004) and a bachelor's degree in economics from Bilkent University (2001). Her research interests include circular economics, applied macroeconomics, financial markets, and economic growth and development.

Natalya Ketenci is professor of economics at Yeditepe University, Turkey, where she has been employed for the past seventeen years. Since 2022, she is the director of the Management Application and Research Center at Yeditepe University (MARC-YUVAM). She received her PhD degree in economics from the University of Reading, UK (2003) and an MSc in Mathematics from the Kazak State National University (1996). At the same time, she is a member of the Centre for Euro-Asian Studies at the University of Reading's Henley Business School. Dr. Ketenci was assisting in the preparation of the Transition report 2001: Energy in transition in the European Bank for Reconstruction and Development, London (UK). Her works have been focused on international trade issues in the European Union and emerging countries. Her current research has a focus on energy issues in Turkey and on energy links between European and transition countries. She is one of the editors and authors of *Economics and Politics of Energy in the Middle East and Eastern Europe* (2021) and *Turkish-Russian Relations: Prospects and Challenges* (2019). Her publications include contributions to such journals as Energy, Russian Journal of Economics, Empirical Economics, Economic Modeling, and Research in Economics.

Duaa Barakat received her bachelors degree in accounting and auditing from Birzeit University in 2009. Following her bachelors degree, she worked as a financial supervisor at Middle East International School in Palestine. During her work at Middle East International School, she received her MA (2014) degree in the field of economics from Birzeit University. In 2015 she moved to Riyadh, Saudi Arabia; during her stay she was employed as a lecturer at Mahara training center. In 2018 she moved to Istanbul, Turkey, then received her PhD between Jan 2018 and December 2022 in the field of economics from Yeditepe University studying under associate professor Dr. Alper Altinanahtar. The working title of her thesis is "Erratic Crude Oil Prices Fluctuations Effects on the Economic Development of Turkey: The Case of Coronavirus Pandemic (Covid-19)." Her research addressed key questions in relation to the role of oil and energy markets on the response of Turkey's economic growth, with COVID-19 as an extreme case. This research is designed to be a valuable academic contribution to research of economic and econometric analysis. In May 2022, she received a certificate from the ICPESS 2022 Sakarya Conference for presenting two articles named "Assessing Short- and Long-Term Effects of COVID-19 Crisis on Stock and Energy Markets in Turkey" and "A Revolutionary Covid-19 Recovery Approach Based on Circular Economy: How Turkey Create a Path for a Sustainable Economy and Healthy Future?" Currently her works has been focused on various economic topics such as Micro and Small Enterprises, oil market, energy market, healthcare, pandemics, and circular economy. She can be contacted at: duaa.barakat@std.yeditepe.edu.tr.

www.ingramcontent.com/pod-product-compliance
Ingram Content Group UK Ltd.
Pitfield, Milton Keynes, MK11 3LW, UK
UKHW041452071026
15587UKWH00010B/790

9 781666 921083